符号中国 SIGNS OF CHINA

中国盆景

CHINESE BONSAI

"符号中国"编写组 ◎ 编著

中央民族大学出版社
China Minzu University Press

图书在版编目(CIP)数据

中国盆景：汉文、英文 /"符号中国"编写组编著. —北京：
中央民族大学出版社, 2024.8
（符号中国）
ISBN 978-7-5660-2304-9

Ⅰ.①中… Ⅱ.①符… Ⅲ.①盆景—观赏园艺—介绍—中国—汉、英 Ⅳ.①S688.1

中国国家版本馆CIP数据核字（2024）第017054号

符号中国：中国盆景 CHINESE BONSAI

编　　著	"符号中国"编写组
策划编辑	沙　平
责任编辑	杜星宇
英文指导	李瑞清
英文编辑	邱　械
美术编辑	曹　娜　郑亚超　洪　涛
出版发行	中央民族大学出版社
	北京市海淀区中关村南大街27号　邮编：100081
	电话：（010）68472815（发行部）　传真：（010）68933757（发行部）
	（010）68932218（总编室）　　　　（010）68932447（办公室）
经销者	全国各地新华书店
印刷厂	北京兴星伟业印刷有限公司
开　　本	787 mm×1092 mm　1/16　印张：10
字　　数	138千字
版　　次	2024年8月第1版　2024年8月第1次印刷
书　　号	ISBN 978-7-5660-2304-9
定　　价	58.00元

版权所有　侵权必究

"符号中国"丛书编委会

唐兰东　巴哈提　杨国华　孟靖朝　赵秀琴

本册编写者

马利琴

前言 Preface

　　盆景艺术是直接受自然界的启迪而产生的。盆景师法自然，是以各种各样的植物、山石、水、土等为素材，经过一定的艺术加工，将天然情景移入咫尺的盆中，所创造出的"第二自然"。盆景被誉为"无声的诗，立体的画"，盆景艺术是集园林栽

The art of bonsai (literally tree in a pot) is directly inspired by nature, which takes various plants, rocks, water, soil, etc. as source materials. Throngh certain artistic treatment, people moves the natural scene into a small pot and thus creates "the second nature". Hailed as silent poetry and three-dimensional painting, bonsai has integrated

培、文学、绘画等艺术于一体,把诗情画意熔于一炉,使人从中获得诗画一般隽永而美妙的艺术享受。

本书不仅介绍了盆景的类别和流派,还对盆景文化的发展渊源、艺术特色、重要价值等作了简要的概括。读者可通过本书,对中国的盆景艺术有一个全方位的了解。

garden cultivation, literature and painting all in one, hence is able to radiate poetic and pictorial splendor that provides viewers profound, pleasant and everlasting aesthetic enjoyment.

This book not only introduces the classification and schools of bonsai, but also touches upon its origin of history, artistic features and great values, through which you may have a comprehensive idea of the Chinese bonsai art.

目录 Contents

盆景概述
Overview of Bonsai 001

历史渊源
Origin of History 002

盆景的艺术特点
Artistic Features of Bonsai 022

盆景类别
Categorization of Bonsai 033

树桩盆景
Tree Stump Bonsai 034

山水盆景
Landscape Bonsai 054

水旱盆景
Water-and-land Bonsai 069

花草盆景
Flower Bonsai 078

微型盆景
Mini Bonsai 088

挂壁盆景
Wall-hanging Bonsai 094

异形盆景
Fancy Bonsai 099

多彩的盆景流派
Colourful Bonsai Genres 101

岭南盆景
Lingnan Style Bonsai 102

川派盆景
Sichuan Style Bonsai 110

扬派盆景
Yangzhou Style Bonsai 115

苏派盆景
Suzhou Style Bonsai 122

通派盆景
Nantong Style Bonsai 129

海派盆景
Shanghai Style Bonsai 135

浙派盆景
Zhejiang Style Bonsai 139

徽派盆景
Huizhou Style Bonsai 143

盆景概述
Overview of Bonsai

　　盆景艺术起源于中国，是中国传统文化中的艺术珍品。历代的盆景制作者用自己的聪明智慧和辛勤汗水制作出各种类型的盆景，使之成为一种内容丰富、形式多样的文化艺术创作活动。

Originated in China, bonsai is an artistic treasure in traditional Chinese culture. Over the ages, cultivators has brought out all kinds of bonsai by wisdom and hard work which further enriches this cultural and artistic creation both in content and form.

> 历史渊源

中国盆景历史悠久，源远流长。在距今八九千年前的新石器时代，人们就已经知道将植物栽入盆中来观赏了。在浙江省余姚市河姆渡新石器遗址中，曾发现了一片刻

> Origin of History

Bonsai enjoys a long history in China. Dating back to the Neolithic Age over 8000 to 9000 years ago, people already began to appreciate potted plants. A pottery shard was unearthed at the Hemudu Site of Yuyao City, Zhejiang Province. It is carved with image of potted plant and hence regarded by some as the earliest evidence of Chinese bonsai.

In the Western Han Dynasty (206 B.C.-25 A.D.) when Zhang Qian the imperial envoy went to the Western

• 紫檀嵌松石玉料石榴盆景（清）
Pomegranate Bonsai Made of Rosewood and Turquoise (Qing Dynasty, 1616-1911)

有盆栽植物的陶片，有人认为这可能是确定盆景起源的最早的证据。

西汉时期（前206—公元25），张骞出使西域时，为了把西域的石榴引种到中原地区来，就采用了盆栽石榴的办法。这也是迄今为止中国最早的关于植物盆栽的记载。

Regions on diplomatic mission, he introduced western pomegranate into the Central Plains of China. For portable convenience, he put it in a pot during the journey. This is so far the earliest written record of potted plant in China.

According to the legend, once there lived a Taoist called Fei Zhangfang who could contract views like mountains,

张骞出使西域

张骞生活在西汉武帝（前140—前87）时期。当时西汉政府与西部边疆的匈奴之间战争不断。汉武帝听说在匈奴的西边有一个叫大月氏的国家与匈奴之间有矛盾，于是就派张骞作为汉朝的使者出访大月氏。张骞在出访途中被匈奴兵发现，他与随行人员全部都做了俘虏，这一关押就是十多年。后来张骞趁匈奴人疏忽逃了出来，历经万难终于回到了长安。不久，他又再一次出使西域。张骞这次出使历时四

- "张骞出使西域"壁画
Mural of Zhang Qian's Journey to the Western Regions

年，终于与大月氏建立了联系。而他经过的路线后来成为中国与西亚交流的重要陆上交通线——丝绸之路。

Zhang Qian's Journey to the Western Regions

Zhang Qian lived during the reign of Emperor Wu of the Western Han Dynasty (140 B.C.-87 B.C.) when constant warfare happened between the government and the Huns at western frontiers. Hearing that the Huns were also in conflict with Dayuezhi, one of Hunnish neighboring countries to the west, Emperor Wu sent Zhang Qian as national envoy to visit Dayuezhi. Unfortunately, Zhang Qian and his followers were discovered by the Huns during the journey and then captured for over a decade. Later when unwatched, Zhang Qian took chance to escape. After all difficulties he finally returned to Chang'an, the capital of the Western Han Dynasty. Soon he went on his second journey to the Western Regions which lasted four years and eventually established contact with Dayuezhi. The path he undertook, later known as the Silk Road, becomes an important traffic line overland for exchange between China and the Western Asia.

据说东汉时期的道士费长房能集各地山川、亭台楼阁、帆船舟车、树木河流于一个盆缶之中，在河北省望都县东汉墓中发现的一幅壁画也表现了这一传说。画面中一个圆盆放置在一个方形几架之上，盆里栽种着六枝红花。这种植物、盆钵、几架三位一体的盆栽形式已经与现在的盆景艺术十分相像，可以说是盆栽植物向盆景艺术的过渡。

rivers, pavilions, sailboats, carriages and trees at all parts of the country into one small pot. The mural excavated at a tomb of the Eastern Han Dynasty (25-220) in Wangdu County, Hebei Province certifies the legend. From the mural we can see a round pot planted with six red flowers on a square table. Plant, pot and table make a three-in-one pattern that quite resembles modern bonsai and indeed foreshadows such a transition from potted plant to bonsai art.

盆景与盆栽

盆景是从盆栽进一步提高发展起来的,盆景与盆栽有根本的区别。盆栽只是将植物种于花盆之中,从而供四时观赏,其审美的对象只在于枝叶、花朵、果实等颜色形状。而盆景,不仅要达到盆栽的观赏目的,还必须通过精心的艺术造型,表现出无穷的诗情画意。盆景寄托了作者的艺术情感,是景致与情感的交融,是自然美与艺术美的有机结合。

Bonsai and Potted Plant

Bonsai is developed from potted plant yet retains fundamental difference. In the latter case, plants are simply raised in the pot so that you can appreciate their foliage, flower, fruit, color and shape over the year. In the former case, however, artists not only need to preserve the ornamental value of potted plants, but reveal infinite poetic and pictorial charm through careful artistic design. Moreover, bonsai sustains the affections of men, merges the view with emotion and exemplifies an organic combination of natural and aesthetic beauty.

- **海棠盆景**
  Chinese Flowering Crabapple Bonsai

南北朝时期（420—589），六朝文人追求山水美的意境，发扬了汉代园林设计中"一池三山"的模式，把大自然引入庭园之中，追求诗情画意，这为唐宋及以后盆景的兴盛打下了良好的基础。

唐代（618—907）国力强盛、文化繁荣，促进了盆景艺术的发展。当时无论是官廷还是民间，制作赏玩盆景都成为一种风尚。制作者们尝试着运用山水画的创作理论将山石与植物组合成盆景，加强了盆景的意境美。在唐代，人们对于盆景的鉴赏能力已经达到了一定程度。在

During the Northern and Southern dynasties (420-589), in pursuit of landscape beauty, literati carried forward the mode of "One Lake, Three Mountains" which was very popular in garden design of the Han Dynasty (206 B.C.-220 A.D.) and thus brought nature into the walled courtyard for poetic dwelling. This laid a good foundation for the prospering of bonsai art in the Tang and Song dynasties, and all later generations.

In the Tang Dynasty (618-907) national and cultural prosperity further advanced the development of the bonsai art. No matter in the royal court or among

• 《六尊者像》卢棱伽（唐）
Six Venerable Masters, by Lu Lengjia (Tang Dynasty, 618-907)

一池三山

在古代传说中，东海里有蓬莱、方丈、瀛洲三座仙山，山上长满了长生不老药。古代帝王都梦想能长生不老，永久统治，于是汉武帝在长安城修建了瑶池（太液池）和"三仙山"。此后，"一池三山"就成为历代皇家园林的传统格局。

One Lake, Three Mountains

Based on the ancient legend, there were three divine mountains in the East China sea, respectively Mt. Penglai, Mt. Fangzhang and Mt. Yingzhou, all covered with elixirs. Emperor Wu of the Western Han Dynasty (156 B.C.-87 B.C.), like all other ancient emperors, dreamt of everlasting life and permanent rule, so he built a Jasper Lake (the adobe of immortals, also known as Taiye Lake) and three divine mountains in Chang'an City. Thenceforward, "One Lake, Three Mountains" becomes the traditional pattern for imperial gardens.

● 北京北海琼岛
Jade Islet at Beihai Park, Beijing

唐代章怀太子李贤墓内有侍女手捧盆景的壁画。画中一位侍女手捧黄色圆盆，盆中有数块拳头大小的石头，上面长着两棵小树，树上结着红绿两色的果子。画中绘的盆景与现在的附石盆景有许多类似之处。由此可见，当时的人们已懂得以石衬树，形成形态、高低、虚实的对比，以此来增添自然气息。而且以盆景作为室内装饰已经成为当时的一种风俗，这也促进了盆景的商品

the common folks, bonsai cultivation and appreciation became a fashion. To reinforce its artistic conception, cultivators managed to make bonsai out of rocks and plants by the creation theory of landscape painting. In the Tang Dynasty, people already had certain discernment to appreciate the bonsai art. For example, at the tomb of Li Xian, formally Crown Prince Zhanghuai of the Tang Dynasty exist murals portraying maids holding bonsai. One maid has a yellow round pot filled with several pieces of fist-sized stones where two small trees grow out with red and green fruit. The painted bonsai shares some similarities with modern bonsai of trees and stones. From this, it can be seen that people at that time already knew how to set off trees by stones, thus creating a contrast in form and height as well as between virtuality and reality to enhance natural flavor. Besides, using bonsai for indoor decoration was quite prevalent at that time which also promoted bonsai

● 唐代章怀太子李贤墓内侍女手捧盆景的壁画
Mural Painted with Maids Holding Bonsai at the Tomb of Li Xian (Crown Prince Zhanghuai of the Tang Dynasty, 618-907)

化。唐代卢棱伽所绘的《六尊者像》中描绘有一人向一位僧人敬献盆景的情景，其中所绘的盆景应为树石盆景。

宋代盆景在继承唐代盆景的基础上进一步发展，树木盆景与山水盆景的区别更加明确。北京故宫博物院里收藏的宋人所绘《十八学士图》中有两轴绘画都绘有苍劲古松、老干虬枝、悬根出土的盆景。这是宋代盆景艺术的一个物证，从中可以看出当时盆景制作技艺的高超。

由于宋人赏石、玩石的情趣更加浓厚，这也使得宋代的山水盆景制作技艺有了显著提高。赵希鹄在《洞天清录集》中分析了各种山石的特点和山水盆景的制作方法。他写道："怪石小而起峰，多有岩岫耸秀，镶嵌之状，可登几案观玩，亦奇物也。色润者固甚可爱，枯燥者不足贵也。道州石亦起峰可爱。川石奇耸，高大可喜，然人力雕刻后，置急水中舂撞之，纳之花栏中，或用烟熏，或染之色，亦能微黑有光，宜作假山。"

此外，宋代有了对盆景的题名

commercialization. For another instance, in *Six Venerable Masters* painted by Lu Lengjia of the Tang Dynasty, we can see a man presenting bonsai of stones and trees to a monk.

Bonsai of the Song Dynasty (960-1279) inherited the legacy of the Tang Dynasty (618-907) yet got further developed on that basis. The distinction between tree bonsai and landscape bonsai became only sharper. At the Palace Museum, there stores a masterpiece titled *Eighteen Scholars* by an unknown painter of the Song Dynasty. Two scrolls are painted with bonsai that features vigorous ancient pines with aged trunk, twisted branches and exposed roots. This serves evidence of the bonsai art in the Song Dynasty, and also highlights the superb craftsmanship of bonsai cultivation at that time.

In the Song Dynasty (960-1279), as people developed strong interest for stones, the accomplishment in landscape bonsai was significantly improved. In Chapter *Queer Rocks Analysis* of *Record of the Pure Registers of the Cavern Heaven*, Zhao Xihu elaborates upon the features of a variety of rocks and also the way to cultivate landscape bonsai. In his words, "The queer rocks are small yet

- 《十八学士图》佚名（宋）
在画面的右下角有一盆松树盆景，盆景中的松枝造型自然美观，叶翠枝茂，根系悬露土面，古朴、苍劲。

Eighteen Scholars, by Anonymous (Song Dynasty, 960-1279)
In the lower right corner of the scroll stands a pine bonsai in leafy profusion. The flourishing twigs extend with natural grace; the roots, exposed on the ground, look vigorous and of primitive simplicity.

之举。如南宋诗人范成大爱玩英德石、灵璧石和太湖石，并在奇石上题"天柱峰""小峨眉""烟江叠嶂"等名称。

元代盆景实现了体量小型化的飞跃，这对盆景的大力普及和推广起到了促进作用。当时的高僧韫上人，周游各国，饱览祖国名川大山

peaked, having elegant towering ridges that look mosaic-like. You can put on the table to appreciate their unique beauty. Those of refined and moist color are in particular lovely whereas the rough and dry ones are not quite valuable. Rocks from Daozhou have lovely peaks, while those from Sichuan Province look lofty and grotesque. Large rocks are quite attractive. After manual carving and torrent battering, they are usually placed within flowery fence, then smoked or dyed, thus attain a slight black hue and may serve as rockery."

In addition, during the Song Dynasty (960-1279) people began to inscribe names on bonsai. Take Fan Chengda, a famous poet of that historical period for instance. So fond of Yingde stone, Lingbi stone and Taihu stone, he would inscribe names like Tianzhu Peak (meaning the pillars of heaven), Mount E'mei the Junior, Misty Rivers and Rolling Peaks on the queer stones.

When it came to the Yuan Dynasty (1206-1368), bonsai made a big leap in miniaturization, which further helped to popularize this remarkable art. At that time there was an eminent monk named Yunshangren who after touring around the country to relish national landscape, then

● 精巧的盆景
Delicate Bonsai

之后，以大自然为基础，打破传统格局，极力提倡盆景的小型化，后人称之为"些子景"。这种"些子景"与现在的中型盆景差不多，与微型盆景有一定的差别。

明清两朝，盆景类别更加多样，除了山水盆景、旱盆景、水旱盆景外，还出现了镶金缀玉的盆景。它们是由黄金、象牙、宝石、景泰蓝等珍贵材质制作而成的，高贵典雅，精巧别致。它们的出现，

based on nature, broke the traditional pattern and strongly advocated small-sized bonsai—later known as *Xiezijing* (literally small scenery). *Xiezijing* looks similar to modern medium-sized bonsai yet a bit different from miniature bonsai.

During the Ming and Qing dynasties (1368-1911) bonsai expanded in variety. Aside from landscape bonsai, rock-filled bonsai, water-and-land bonsai, there also emerged bonsai inlaid with gold and jade. They are made of gold, ivory, gems,

● 掐丝珐琅"玉堂富贵"盆景（清）
Enamel Filigreed Bonsai Named "Jade Chinese Crabapple Brings Wealth" (Qing Dynasty, 1616-1911)

• 银嵌金累丝穿珠梅花盆景（清）
Bead Shaped Plum Blossom Bonsai Made of Silver Inlaid with Gold Filigree (Qing Dynasty, 1616-1911)

cloisonné and other rarities, hence elegant and exquisite. Their appearance further enriches the variety of ancient bonsai.

During the Ming and Qing dynasties (1368-1911), not only did bonsai craftsmanship get matured, but many relevant books came into being which have systematically discussed all theories related to tree species, stone types, production and even placement of bonsai. It is during this period that the Chinese bonsai art achieves a big theoretical leap. In *Other Things in Moral Cultivation* written by Tu Long of the Ming Dynasty (1368-1644), he said that, "Good bonsai are small enough to be put on tables; otherwise they should be grown in the courtyard." In addition to application and positioning, he also proposed that bonsai cultivators should turn to the ancient trees portrayed by traditional painters for reference. As to the trunk coiling technique, he insisted on following nature, thus though artificially cultivated, the bonsai may appear naturally formed. Li Dou of the Qing

进一步丰富了古代盆景的类型。

明清时期，不但盆景技艺已经趋于成熟，而且还出现了许多的盆景专著，对盆景的树种、石品、制作、摆置等作了理论上的系统论述。中国盆景艺术的理论在这一时期得到了飞跃和升华。明代屠隆在《考槃余事》一书中写道："盆景以几案可置者为佳，其次则列之庭榭中物也。"介绍了盆景的应用配置，同时他还提出应以古代画家笔下的古树为参照对象进行创作。对于树桩的蟠扎技艺，他主张师法自然，强调虽由人作，宛自天开。清代李斗所著的《扬州画舫录》中提到乾隆年间，扬州已有花树点景和山水点景的创作。而陈扶摇的《花镜》中专门设有《种盆取景法》一节，论述盆景用树的特点和经验。

Dynasty once mentioned in *Travelling Records of Yangzhou* that early in the Qianlong Period (1736-1795), people in Yangzhou already began to use flower, tree or landscape for decoration. In his famous book *Flower Mirror*, Chen Fuyao devoted one special chapter titled *View Selecting Technique of Potted Plant* to discuss the characteristics and experience of tree disposition in bonsai.

- 《松友图》（局部）杜琼（明）

  画作展现的是庭院之景：竹篱环绕庭院，房前山峦叠翠，泉水曲径，有数人活动其间。而在画面中放置于石桌上的两盆松树盆景更增添了主人的高洁品质。

  Pines (Partial), by Du Qiong (Ming Dynasty, 1368-1644)

  The painting depicts a scene of the courtyard which is bamboo-fenced with verdant mountains rolling in front, stream winding by and some people moving about. Two pine bonsai placed on the stone table further brings out the noble quality of the courtyard owner.

盆景植物中的"四大家""七贤""十八学士"和"花草四雅"
Four Masters, Seven Sages, Eighteen Scholars and Four Elegant Plants in the Bonsai World

盆景制作者根据人们对某种观赏植物的喜爱和推崇,往往给其以"贤人""学士"的雅号,于是便有了盆景植物中的"四大家""七贤""十八学士"和"花草四雅"。

Based on people's admiration to certain ornamental plants, bonsai cultivators often give them esteemed names like "wise man" or "scholar". Hence now we have Four Masters, Seven Sages, Eighteen Scholars and Four Elegant Plants.

四大家:金雀、黄杨、迎春、绒针柏。

Four Masters refer to cytisus, boxwood, winter jasmine and velvet-needled cypress.

- **黄杨盆景**

黄杨是黄杨科黄杨属植物,常绿小乔木或灌木状。树干灰白光洁,枝条密生,枝四棱形。叶对生,革质,全缘,椭圆或倒卵形,表面亮绿色,背面黄绿色。花簇生叶腋或枝端,4—5月开放,颜色呈黄绿色。果实呈卵圆形。该属尚有雀舌黄杨、珍珠黄杨。黄杨生长缓慢,寿命长,株矮小,耐修剪,是制作盆景的良好树种。

Boxwood Bonsai

Boxwood is of the genus *Buxus* in the family Buxaceae. The trunk is gray and spotless with dense prismatic branches. The leaves are opposite, leathery, entire, oval or obovate in shape, bright green in the front side and yellow-green in the back. Flowers usually unfold between April and May, yellow-green in color and some cluster around the leaf axils, some at ends of the twigs. Fruit appears oval in shape. Now we still have *Buxus bodinieri* and *Buxus sinica* var. *parvifolia* M.Cheng of its genus. Since boxwood is slow-growing, long-lived, dwarf-stemmed and pruning resistant, it is an excellent tree species for bonsai.

七贤：黄山松、缨络柏、榆、枫、冬青、银杏、雀梅。

Seven Sages refer to *Pinus hwangshanensis* W.Y.Hsia, *Juniperus communis* L., elm, maple, holly, ginkgo and *Sageretia Thea*.

- **银杏**

银杏属于落叶乔木，叶片扇形，有两叉状叶脉，顶端常二裂，有长柄。雌雄异株，球花生于短枝顶的叶腋或苞腋，花期4—5月，种子核果状，椭圆形至近球形，外种皮肉质，有白粉。银杏生命力强，叶形奇特，易于嫁接繁殖和整形修剪，是制作盆景的优质树种。

Ginkgo

Ginkgo is a deciduous and dioecious tree. Long-stemmed and fan-shaped, the leaves are unique with two forked veins radiating out into the blade and often bifurcate at the tip. It usually flowers during April and May. The cones sprout from the axils or axillary buds at the ends of the short shoots. The seed is drupe-like, oval or nearly spherical in shape. Its fleshy outer layer is covered with white powder. Ginkgo features strong vitality and peculiar leaves. Easy to graft, propagate and prune, it is therefore another excellent tree species for bonsai making.

十八学士：梅、桃、虎刺、吉庆果、枸杞、杜鹃、翠柏、木瓜、腊梅、南天竹、山茶、罗汉松、西府海棠、凤尾竹、石榴、紫薇、六月雪、栀子。

Eighteen Scholars refer to plum, peach, *Damnacamthus indicus*, *Solanum pseudo capsicum* L., Chinese wolfberry, rhododendron, Chinese incense cedar, Chinese quince, wintersweet, *Nandina domestica* Thunb., camellia, *Podocarpus macrophyllus*, *Malus micromalus*, fernleaf hedge bamboo, pomegranate, crape myrtle, snowrose and gardenia.

- 杜鹃

杜鹃又名映山红，为杜鹃花科杜鹃花属常绿或落叶灌木，植株多分枝，枝条细而直，有棕色或褐色糙毛。单叶互生，叶片卵状椭圆至披针形，先端尖或稍钝，表面疏生棕色硬毛。花数朵簇生枝条顶端，花冠漏斗形，单瓣或重瓣，花色丰富，有白、粉、红、紫以及各种复色。杜鹃花的品种很多，大致可分为春鹃、夏鹃、春夏鹃、西洋鹃四大类。用其制作的盆景具有根干苍劲古朴，叶片细小稠密，叶色碧绿，耐修剪，花色繁多等特点。

Rhododendron

Rhododendron, also known as azalea, is any of numerous usually evergreen or deciduous shrubs of the genus *Rhododendron*. The plant usually has many branches which are thin yet straight with brown and rough hair. The leaves are alternate and single. The blade can be ovate, oval or needle-shaped with pointed or slightly blunt tip and brown hard hair sparsely grown on the surface. Flowers usually cluster at ends of the twigs; the corolla is funnel-shaped, single or double-lobed and varying from white, pink, red, purple to some secondary color. Rhododendron comes in a great many varieties which can be broadly divided into four categories, respectively spring rhododendron, summer rhododendron, spring and summer rhododendron and western rhododendron. Rhododendron bonsai usually features vigorous roots, small, dense and verdant leaves, trimming resistance and a wide range of color.

- **罗汉松**

  罗汉松属罗汉松科，罗汉松属。常绿小乔木，树皮暗灰色，鳞片状开裂；主干挺直，枝条平展而密生。叶螺旋状互生，条状披针形，两面中肋隆起，表面浓绿色，背面黄绿色，有时具白粉。4—5月开花。种子核果状，卵圆形，熟时呈紫红色，似头状，种托似袈裟，全形如披袈裟的罗汉，故名"罗汉松"。

  Podocarpus Macrophyllus

  Podocarpus macrophyllus is a small evergreen tree in the genus *Podocarpus*, family Podocarpaceae. The bark is dark gray in color with cracking scales. The trunk stands upright and dense branches extend out. Leaves spirally alternate each other. They are strap-and-needle shaped with swollen midribs, dark green on the surface and yellow-green in the underside, sometimes having white powder. The flowering period goes between April and May. The drupe-like seed is oval-shaped. When ripened, it turns purple-red in color like a human head and the seed aril looks just like a Buddhist wearing robe. As it resembles the Arhat so much, podocarpus is also known as Arhat Pine.

- **石榴**（图片提供：FOTOE）

  石榴属石榴科，石榴属。落叶灌木或小乔木，枝顶端多为棘刺状。单叶对生或簇生，长椭圆形或倒披针形，表面有光泽，新叶呈红色。花通常一至数朵顶生，有大红、粉红、黄、白诸色，花瓣皱缩，萼紫色。花期5—7月，盛开时红艳似火，霞光照眼。果球形，红黄色，顶端有宿存萼片。石榴挂果期长，绿叶红果，是制作盆景的重要树种。

  ### Pomegranate

  Pomegranate is a deciduous shrub or small tree of the genus *Punicagranatum*, having thorny tipped branches. The leaves are simple, opposite or clustered, oblong or oblanceolate, glossy in surface; the young ones are usually red in color. Flowers usually blossom singly or in small clusters at the ends of the twigs, ranging from red, pink, yellow to white in color. The petals are wrinkled and the calyxes are purple. The flowering period falls between May to July. When in blossom, it will look as red as fire and shining like rays of sunshine. The fruit is spherical, red and yellow in color, having persistent sepals. Pomegranate has a long fruiting period. With green leaves and red berries, it is an important tree species for bonsai making.

- 栀子

栀子属茜草科，栀子属。常绿灌木，树冠圆球形，枝丛生。单叶对生，革质，长椭圆形，全缘，表面浓绿色，有光泽。花冠呈高脚碟状，色洁白，腋生或顶生，花萼圆筒形，先端6裂，5—7月开花，芳香扑鼻。果实卵形，有6纵棱，萼宿存，11月成熟，橙黄色。

Gardenia

Gardenia belongs to the genus *Gardenia* in family Rubiaceae. It is an evergreen shrub marked with spherical crown and tufted branches. The leaves are simple, opposite, leathery, oblong, entire, glossy and dark green in appearance. The corolla looks like a high-stemmed dish, white in color, and sprouts from axillary or terminal buds. The calyx is cylindrical with six lobes. When it comes into blossom from May to July, pleasant aroma will flow everywhere. The fruit is ovate with six vertical ribs and persistent sepals. It usually ripens in November and is orange in color.

花草四雅：兰、菊、水仙、菖蒲。

Four Elegant Plants refer to orchid, chrysanthemum, narcissus and calamus.

- 菊

Chrysanthemum

中华人民共和国成立之后，国家对盆景艺术采取了保护、发展和提高的政策，使古老的盆景艺术获得了新生。1981年，中国花卉盆景协会在北京成立。1988年，中国盆景艺术家协会在北京成立。

1979年10月，中国盆景首次参加了在联邦德国首都波恩举办的第15届国际园艺博览会，因一举夺得各种奖牌13枚而轰动整个欧洲。其后，中国盆景先后在十余个国家展出，均受到好评。通过对外展出与广泛交往，中国盆景艺术蜚声海外。

After the founding of the People's Republic of China, policies had been made to protect and further advance the bonsai cultivation and thus revived the ancient art. In 1981 and 1988, China Flower Bonsai Association and China Bonsai Artist Association were successively set up in Beijing.

In October 1979, Chinese bonsai for the first time participated in the 15th International Horticultural Exposition held in Bonn, the former capital of the Federal Republic of Germany. China won thirteen medals which made a great sensation over Europe.

Thereafter, Chinese bonsai was successively displayed in more than 10 countries and all well-received. Through exhibition and wide exchange, Chinese bonsai now enjoys a high reputation overseas.

- 对节白蜡盆景
  Fraxinus Hupehensis Bonsai

> 盆景的艺术特点

盆景是以植物和山石为基本材料，在盆内展现自然景观的一种艺术。概括起来，它具有以下特点：

第一，小中见大。盆景是把自然景物缩小到咫尺的盆钵之中，造型具有很强的概括性，能够更加集

> Artistic Features of Bonsai

Bonsai is an art that takes plants and rocks as basic material to set out a natural scene within a pot. In general, its characteristics can be summed up as follows.

First, multum in parvo. That is, bonsai contracts the broad natural scene into a small pot and then by general

• 盆景——云壑松风
Bonsai — Misty Valley and Pines in Wind

中地将大自然的风光面貌表现出来。只要在盆中放入少许水，再巧妙地布置几块山石，就能显示出浩瀚的水域和波澜起伏的峰峦；即使是一棵不足一尺（约33厘米）的树木，经过修剪、矮化、蟠扎等技术也能够具有虬曲苍古的风姿。这便是盆景艺术"以小见大"的魅力所在，也是盆景艺术创作者们追求的理想境界。

第二，活的艺术品。树木花草是盆景造型的主要材质，可以不断生长发育，因此，盆景的艺术性是可变的。随着季节的更替，春花、夏绿、秋果、冬姿等规律性的四季景观可以在盆景上一一展现出来。这种生机盎然的盆景，具有独特的艺术之美。

第三，形式多样。随着时代的发展，生活内容的逐渐丰富，盆景艺术也更趋于多样化，盆景种类也日益增多。中国盆景植物从传统的几十个树种，发展到现在的上百个树种，在树木、山水两大类别的基础上，又创新了水旱、微型、花草、果树等类别。此外，盆景的风格也不断增多，艺师巧匠们大胆进行艺术构思，采用独特的艺术表现

modeling, represents the natural scenery in a more intensified manner. All you need to do is to put a little water in a pot, artfully place several rocks around, and then vast waters and rolling ridges seem to unfold before your eyes. Even if the tree is no higher than one *Chi* (a Chinese unit of length, 1 *Chi* equals to 0.33 m), after trimming, dwarfing, coiling and other techniques it will turn out gnarled yet vigorous, twisted but full of grace. Multum in parvo is the exact charm of bonsai art and also the ideal state that artists all aspire for.

Secondly, bonsai is a living work of art. Trees, flowers and grass are the main materials for bonsai styling. They carry on their natural growth, so the artistry of bonsai is also mutable. As the seasons change, regular seasonal views like spring flower, summer verdure, autumn fruit and wintry scenery can all be reflected in the potted plant. Indeed, it is this vitality that contributes to the unique beauty of bonsai.

Thirdly, bonsai has various styles. As the times progress, life also becomes colorful. Bonsai likewise tends to be more diversified and increased in species. In ancient times, Chinese bonsai plants has only dozens of tree species, but today

一景二盆三架

俗语讲："一景二盆三架。"上好的景要放在合适的盆中，还要配以恰当的几架，这样景、盆、架三位一体，才能组合出一件完美的盆景来。

盆不仅是景物的容器，也是具有观赏价值的艺术品，要使盆与景相辅相成就要选好盆。选择盆景用盆时需要注意：盆与景要大小适中，色彩协调，质地相宜，形态相称。一般来讲，树桩盆景多用色浅、口深的紫砂陶盆或深色釉盆；山水盆景则多用色浅、口浅的大理石盆或釉盆。

不同的盆景需要配置不同的几架。常见的盆景几架主要为木质，使用的木料多为上等材质的紫檀木、红木、花梨木、楠木、枣木和黄杨木等。几架的款式有：圆几、方几、长方几、书卷几、角架和博古架。一般来讲，大型的山水盆景多用长方几，微型的盆景可以放在博古架上。

- **石盆**

石盆用汉白玉、大理石或花岗岩等质地较软的石料雕凿而成，一般用于山水盆景或挂壁盆景。

Stone Basin
Stone Basin is carved from white marble, marble, granite and other stones of relatively soft texture. It is usually used for landscape shallow bonsai or wall-hanging bonsai.

View Comes First, Followed by Pot and Table

As the proverb goes, "To make a good bonsai, view comes first, followed by pot and table." That is, first we should select the best natural view, then place it in a fitting pot and finally set it on a suitable table. Only such a trinity (view, pot and table) can make a perfect bonsai.

Pot is not just a container for the natural scene, but also an ornamental work of art. Pot and the natural view should supplement and set off each other. In selection of the right pots, we should pay particular attention to its congeniality to the natural view in terms of size, color, texture and form. Generally speaking, tree stump bonsai mostly use light-colored, deep-mouthed purple clay pot or glazed ceramic pot of dark color, whereas landscape bonsai often use light-colored, shallow-mouthed marble or glazed ceramic pot.

Besides, different bonsai should be placed on different kinds of tables. The common bonsai tables are primarily made of high-quality wood such as red sandalwood, mahogany, rosewood, nanmu wood, jujube wood, boxwood, etc. Its style ranges from round table, square table, rectangular table, to table like unfolded volume, corner bracket, antique-and-curio shelves. Generally, large landscape bonsai will use rectangular table, whereas the miniature ones can be put on antique-and-curio shelves.

- **釉陶盆**

  釉陶盆是用优质黏土在制成的盆坯外面涂上低温釉彩，在温度为900—1200℃的炉内烧制而成。釉彩的色泽很丰富，适宜栽植树木盆景。其中盆的内外均不上釉者被称为"素身盆"。釉陶盆主要产于广东佛山、江苏宜兴等地。

Glazed Pottery Pot

Glazed pottery pot is made of high-quality clay, then coated with low-temperature glaze outside, and finally baked under high temperature about 900℃-1200℃. The glaze is rich in color and hence suitable for tree bonsai. Those unglazed in both outside and inside are called "plain-bodied pot". Glazed pottery pot is mainly produced in Foshan City of Guangdong Province, Yixing County of Jiangsu Province, etc.

● 瓷盆

瓷盆是采用高岭土在1300—1400℃的炉中烧制而成的，质地较细，且坚硬。盆的形状有方形、六角形、八角形、圆形等，在盆的外壁一般都绘制有人物、山水、花鸟、诗词等彩绘图案。瓷盆的吸水透气性差，一般不宜直接栽培植物，多作为套盆。其中以江西景德镇出产的瓷盆最为著名。

Porcelain Basin

Porcelain basin is made of kaolin and then fired in furnace of 1300℃-1400℃, hence fine and hard in texture. Varied in shape, the basin can be square, hexagonal, octagonal, round, etc. The outer side of the basin wall is usually painted with figure, landscape, flower and bird, poetry or other colored designs. However as it features poor drainability and air permeability, the basin is not suitable to cultivate plants and hence mostly used as flowerpot cover. The most famous porcelain basin is produced in Jingdezhen County of Jiangxi Province.

● 紫砂盆

紫砂盆是用江苏宜兴特产的一种优质陶土，在1000—1250℃的高温中烧制而成。紫砂盆内外不施釉，质地细密、坚韧，物理性能良好，排水透气性非常适宜树木花卉生长。制盆巧匠将盆色、形制、款识、题铭、书画、雕刻等技艺融为一体，使紫砂盆在淳朴中见意蕴，给人以视觉的享受。

Purple Clay Pot

Purple clay pot is made of high-quality clay produced in Yixing, Jiangsu Province and then baked under high temperature about 1000℃-1250℃. It is unglazed both inside and outside, fine and tough in texture, and having good physical properties. Fine drainability and air permeability make it fit for the growth of trees and flowers. The pot maker artfully integrates color, form, inscription, epigraph, painting, calligraphy and carving all in one, rendering the purple clay pot at once simple and profound that greatly enhances our visual enjoyment.

● 书卷几
Table-like Unfolded Volume

● 圆几与四角架
Round Table and Corner Bracket

● 博古架（图片提供：FOTOE）
Antique-and-curio Shelf

○ 高方几
High Square Table

盆景——凌空叠翠
Bonsai — Piled Verdure Aloft in the Air

手法，使盆景造型更加千变万化。

第四，意境深远。人们在欣赏盆景的时候，不仅是欣赏盆景的静态美，往往还会触景生情，生发出丰富的联想，从而领略到景外的意境。盆景制作者往往会通过主客、高低、大小、远近、疏密、繁简、虚实、藏露、刚柔、巧拙等形式进行对比，实现"平中求奇"和"不似之似"的境界，这也是盆景独具的艺术魅力。

it already expands to over one hundred species. In addition to tree bonsai and landscape bonsai, it also innovates other categories like water-and-land bonsai, mini bonsai, flower bonsai, fruit tree bonsai, etc. Besides, bonsai style also grows in number. Through bold artistic conception and unique technique of expression, bonsai artists have greatly enriched bonsai styling.

Fourthly, the artistic conception is far-reaching. When appreciating bonsai,

● 盆景——浓荫深处
Bonsai — Deep into the Dense Shade

people not only admire its static beauty, but also are moved by the sight. Hence by fertile imagination, they will associate it with something outside or beyond. Bonsai makers often create contrast between the subject and the object, the simple and the complicated, the void and the substantial, the hidden and the revealed, the hard and the soft, the skillful and the awkward, or in height, size, distance, density, etc. to achieve a particular effect that embeds the extraordinary in the ordinary or resemblance in non-resemblance. This is also the unique charm of bonsai art.

● 盆景——过桥何处是朝阳
Bonsai — Where Is the Morning Sun after Crossing the Bridge

盆景与绘画

盆景被誉为"立体的画",许多绘画的理论在盆景的创作中都得到了充分的体现。例如,绘画讲究立意,"意在笔先",在下笔之前一定要考虑好,而盆景也讲立意当先。绘画讲究构图的比例对称,而盆景在造型上也注重对称、均衡。绘画作品一定要形神兼备,而盆景作品也要做到景意并存。正是由于两者的密切结合才有了众多精美的盆景作品的出现。

Bonsai and Painting

Bonsai is also known as three-dimensional painting, because many painting theories are fully reflected in bonsai making. For instance, painting pays particular attention to conception, stressing that conception goes before all brushwork, which is the same case with bonsai. Similar to painting that is particular about the symmetry of composition, bonsai styling also lays emphasis upon symmetry and balance. As painting manages to unite the form with the spirit, bonsai alike strives for co-existence of view and conception. It is due to the close integration of the two that many exquisite bonsai works come into being.

- 盆景——洪波破云

Bonsai — Big Waves Break the Cloud

盆景类别
Categorization of Bonsai

中国盆景有一个庞大的体系，树桩盆景和山水盆景是其中主要的两种类别。此外，还有水旱盆景、花草盆景、微型盆景、壁挂盆景和异形盆景等类别。不同类别的盆景有着不同的特点。树桩盆景能表现出自然界旺盛的生命力；山水盆景能够将高峡飞瀑浓缩到浅盆中；水旱盆景更接近于山水画；微型盆景造型玲珑别致；挂壁盆景形式新颖，美化居室的效果极佳。

Chinese Bonsai has developed a huge system. Tree stump bonsai and landscape bonsai are the two main types. In addition, there are other types including water-and-land bonsai, flower bonsai, mini bonsai, wall-hanging bonsai and fancy bonsai. All types have different characteristics: tree stump bonsai displays the exceptional vitality of nature; landscape bonsai contains towering cliffs and great waterfalls in a shallow pot; water-and-land bonsai is more close to a landscape painting; mini bonsai is styled in an exquisite and unique way; the original wall-hanging bonsai is usually employed as indoor decoration.

> 树桩盆景

树桩盆景，简称"桩景"，它是以木本植物为主体，以山石、人物、鸟兽等作为陪衬，通过蟠扎、修剪、整形等一系列的技术加工，在盆中表现出旷野巨木或茂密森林

> Tree Stump Bonsai

Tree stump bonsai, also called miniature gardening, employs woody plant as its main part, with rocks, figurines of people, birds and animals as decorations. It exhibits part of the natural scenery in a container, such as a towering tree in the

- 直干式树桩盆景

直干式树桩盆景通常是主干直立。树干一旦长到一定高度，就要进行摘心，达到层次分明、疏密有致的效果。

Tree Stump Bonsai of Straight Trunk Style

In most cases, straight trunk style is characterized by a straight and upright trunk. Once the tree grows to a certain height, it needs to be pinched to give its branches a fine structure and uneven density.

- **斜干式树桩盆景**

  斜干式树桩盆景的树干向一侧倾斜，一般略带弯曲，枝条平展于盆外，整个造型显得险而稳固，生动传神。

  Tree Stump Bonsai of Slanting Trunk Style

  The slanting style trunk tilts to one side and slightly curves, with its branches open flat out of the pot. Overall, it seems to be shaped in a most precarious fashion yet it is very steadily positioned, which is therefore styled very vividly.

等景象的一种盆景。

根据所用树木的种类和特性以及设计制作的特点，树桩盆景分为七种形式：直干式、卧干式、斜干式、曲干式、悬崖式、附石式、垂枝式。

wilderness or a dense forest, by adopting a series of techniques including wiring, pruning and trimming.

According to the types and properties of bonsai trees, as well as the design and workmanship, there are seven types of tree stump bonsai, including straight trunk style, laid trunk style, slanting trunk style, curved trunk style, cliff style, root-over-rock style, and weeping style.

- 卧干式树桩盆景

卧干式树桩盆景的主干通常是横卧着，具有古朴优雅的风度，疏密有致，古怪苍老，野趣十足。

Tree Stump Bonsai of Laid Trunk Style

The laid trunk style is mostly characterized by a horizontally laid trunk with a simple and unsophisticated elegance. The branches are carefully arranged in different density in an odd yet vigorous way as a visual delight of natural view.

- 曲干式树桩盆景

曲干式树桩盆景的主干弯曲，这里的"曲"不是人为的弯曲，而是指曲度变化较大，但又符合自然生长规律的弯曲。常见的是外形上像"之"字的"三曲式"。

Tree Stump Bonsai of Curved Trunk Style

The curved trunk is bent into a series of curves. Those curves are not man-made curves. Instead, the trunk has a great degree of curvature yet still in accordance with the law of nature. The "three-curve style" growing roughly in the shape of a Chinese character "之" is the most common.

- 悬崖式树桩盆景

悬崖式树桩盆景的主干倾于盆外，树冠下垂倒悬。其造型奇险，创作手法灵活多变，最能引起观赏者的共鸣，也最能表现作品踞崖悬生的意境，蕴含努力拼搏的精神。

Tree Stump Bonsai of Cliff Style

The cliff-style trunk stretches out of the pot and then bends downward, with the tree crown hanging down in the air. It can be styled in a fancy and precarious way through flexible and varied techniques, which easily strikes such a chord with its viewers. It also best presents the concept that although being born in an adverse environment, for instance a cliff, the tree still fights to survive and grow vigorously.

- **附石式树桩盆景**

  附石式树桩盆景，也称"树石式"盆景，是将树木与山石巧妙结合为一体的一种盆景形式。其特点是树木栽种在山石上，树根或扎在石缝中，或抱石而生。其风格多样，有的清秀典雅，有的雄浑大气，有的古朴苍劲，有的险峻陡峭，有的开阔壮观。

  Tree Stump Bonsai of Root-over-rock Style

  The tree stump bonsai of root-over-rock style, also called tree-and-rock style, is a bonsai form which smartly combines bonsai trees with rocks. It is characterized by a tree planted directly on a mountain stone, with its roots either growing in the stone cracks or being wrapped around a stone. It can be styled in diverse manners: some are delicate and elegant; some are simple and forceful; some are dangerously steep; some are grand and powerful.

盆景的室内摆放

室内摆放的盆景以中小型或是微型盆景为主。摆放时需要注意以下三点：

第一，盆景大小与室内空间要协调。如果在较小的室内空间里摆设大型的盆景，会给人一种压抑感；而在较大的室内空间放置微型盆景也会让人感觉不相协调。一般来讲，客厅可以布置略显壮丽的中型山水盆景；角落处适宜放置落地高几架，上设悬崖式树桩盆景；茶几、书桌上适合摆放小型盆景。

第二，盆景的色彩要与室内整体的格调相协调。如果房间的格调是简洁素雅的，那么在盆景的选择上也要倾向于朴实无华。如果房间的格调是热烈的，那么就

- **室内摆放的盆景**（图片提供：全景正片）
Bonsai for Indoor Display

- 室内摆放的盆景（图片提供：全景正片）
  Bonsai for Indoor Display

要选择一些颜色鲜艳的盆景来装点。

第三，盆景摆放的位置也要适宜。树桩盆景、花草盆景要放置在光线能照射到的地方，以利于其生长。此外，盆景摆放的高度也影响欣赏的效果。如果山水盆景的摆放位置略低于人的视线，那么就能够俯瞰到整座盆景中的风光。

Indoor Display of Bonsai

Bonsai displayed indoors are mainly small and medium size bonsai or mini bonsai. The following should be noted when displaying bonsai indoors:

Firstly, the size of bonsai should be harmonious with the indoor space. Large bonsai pots in a small room will make people feel oppressive, while mini bonsai placed in a large room will also be considered inharmonious. In general, landscape bonsai of medium size is placed in living room; tall spider-leg stands with tree stump bonsai of cliff style on display will fit the corner space; small-sized bonsai best fits tea tables and desks.

Secondly, the color of bonsai ought to be harmonious with the general room style. If the room has a simple, quiet and elegant style, the selected bonsai needs to be plain and unadorned. However, if the room has an ornate style, bonsai with bright and rich colors tend to be selected.

Thirdly, the bonsai also needs to be appropriately positioned. Tree stump bonsai and flower bonsai should be kept in places where the sunshine reaches, which would be beneficial for bonsai plants to grow. In addition, at which height the bonsai is placed also affects appreciation. If landscape bonsai is placed slightly lower than the viewer's line of sight, the viewer can overlook the whole landscape the bonsai exhibits.

树桩盆景的制作

制作树桩盆景首先要选择树桩素材。一般选择枝叶细小、生长慢、寿命长、易成活、耐修剪的树木，如枫树、黄杨、榆树、榕树、茶树等。

树桩的获得主要有两种方式，

Creation of Tree Stump Bonsai

In terms of the creation of tree stump bonsai, the first step is selecting stump as its source material. Generally trees with small twigs and leaves, a slow growth rate, a long life span, great viability as well as a better ability to tolerate regular pruning tend to be selected. Those trees

一种是野外挖掘，另一种是人工栽培。野外挖掘树桩一般选择春季树木萌芽时节或是秋季树木落叶时节，以提高所取树桩的成活率。人工栽培树桩在其生长期间就要做好修剪工作，以减少盆景制作过程中的工序。

include maple, Chinese boxwood, Chinese elms, Chinese banyans, tea trees, etc.

There are two ways to obtain a stump specimen: one is collection in the wild and the other method is artificial cultivation. Collection of stumps in the wild is usually conducted in spring, when trees begin to sprout, or in autumn when leaves begin to fall. Those two seasons are chosen for a higher rate of viability of the collected stumps. Stumps cultivated artificially ought to be kept in a state of well pruned during its growth period, so that the processes can be simplified when those stumps are employed for bonsai creation.

- **榆树盆景**（图片提供：FOTOE）
  榆树属榆科，榆属，落叶乔木。树冠圆球形。树皮灰黑色，纵裂而粗糙。小枝灰色，常排列成二列状。叶椭圆状卵形，先端尖，基部稍歪，边缘具单锯齿。花期3—4月，紫褐色，簇生于一年生枝上。翅果近圆形或倒卵形，先端有缺裂。种子位于翅果中央，4—5月果熟。

Chinese Elm (*Ulmus parvifolia*) Bonsai

Chinese Elm (*Ulmus parvifolia*) belongs to the genus *Ulmus*, family Ulmaceae. It is a deciduous tree with a spherical crown. Its grey-black bark is cracking into vertical strips and coarse. Its grey twigs are in opposite pairs and its leaves are elliptical-ovoid shaped with acuminate apexes, slightly oblique basal parts and serrated margins. The blooming period is from March to April and the flowers produced on annual shoots are puce and fasciculate. The samara is near round or obovate, with the apex dehiscent and a seed positioned in the centre of each fruit. The fruit mature period is from April to May.

- **榕树盆景**（图片提供：FOTOE）

榕树属桑科，榕属，常绿乔木，树冠庞大，呈广卵形或伞状；树皮灰褐色，枝叶稠密。叶革质，椭圆形或卵状椭圆形，有时呈倒卵形，长4—10厘米，全缘或浅波状，单叶互生；叶面深绿色，有光泽。聚花果腋生，近球形，颜色初为乳白色，成熟后为黄色或淡红色，花期5—6月，果期9—10月。

Chinese Banyan (*Ficus microcarpa*) Bonsai

Chinese Banyan (*Ficus microcarpa*) belongs to the genus *Ficus* and family Moraceae. It is an evergreen tree with a large crown that is broadly ovoid or umbrella shaped. It has taupe bark and dense foliage. Its leaves are leather-like, elliptical or ovoid-elliptical and sometimes obovate, with the length of 4 cm-10 cm. The leaves are either full margin or undulated univalent leaves that are growing mutually. Its foliage is dark green with a glossy surface and its axillary compound fruit is near spherical in lacte or faint red. The blooming period is from May to June and the fruit mature period is from September to October.

树桩的选择

通常情况下，树木都生长在野外，在环境因素的影响下，会生得各式各样，因此，制作盆景的时候要对树桩的头、根、干、枝做全面的鉴别。整株树的比例、树的基部、树根与树干的造型等要配合得当。如果树干挺拔雄伟，树头就不能有偏根，得有板根或三面露根。如果树干是悬崖形状的，就要有同方向自然生长的偏根。其次，树木的基部要三边或四边长根，裸露生长在土层上面。根的直径最好在一厘米左右，太小缺乏自然美，太大不易生长。树根的走向要从中心辐射开来，像风车一样向不同的方向伸展。此外，每株树都有自己的根系，但树与树之间又是由同一母体相联结的，这种树头叫丛林式树头，是一种十分难得的树形。

- **树桩盆景**
  Tree Stump Bonsai

Selection of Stump

In general, trees grow in the wilderness and they are shaped differently by the surrounding environment. Therefore, before creating a tree stump bonsai, one needs to carry out a comprehensive identification of the stump base, roots, trunk and branches, including the proportion of the tree as a whole, as well as whether the styles of the tree base, roots and trunk are harmoniously corresponding with each other. If the trunk is upright and imposing, the basal trunk is required to have no one-sided root but buttress roots or three well-spaced roots visible from the front. If the trunk is cliff style, it is required to have natural one-sided roots, the side of which is the direction of the slant in the trunk. Next, the basal trunk is required to have longer roots on its three sides or all around, growing above the soil and being exposed to the air. The diameter of its roots is ideally around 1 cm. Roots too small have no natural beauty, while roots too large are difficult to survive. The roots extend from the basal trunk and radiate outward in all directions like blades of a windmill. In addition, each tree has its own root system and all together they are attached to the parent tree's roots. This type of basal trunk is called forest-like style, which is a very rare shape.

第二是造型，就是在树桩原有的形态基础上进行再创作，使其具有某种艺术形象。对树桩进行造型的方法主要有蟠扎和修剪两种。蟠扎，是对树桩的枝干进行适当弯曲，使其产生一定的弧度。修剪，是对蟠扎后的树桩进行处理，使其变得疏密得当，造型更加完美。

第三是布局，通常来说，树桩盆景的制作者会根据材料的实形来构图。每种树桩的本质和形态都是

The second step is styling, which refers to the artistic creation on the basis of the original shape of the stump by imbuing it with certain artistic qualities. The two main methods to style a stump are wiring and pruning. Wiring is to bend the trunk and branches of the stump to make proper curves. Pruning is processing the stump after wiring to give its branches an uneven density and perfect its style.

The third step is composing bonsai arrangement. Generally speaking, tree stump bonsai creators will compose

不同的，加工时要顺其自然，力求简单，不能强拉硬拽地对其进行改造。如果在制作时，硬要将一种树坯改造成大树形、飘斜形、蟠曲形盆景，不但会违反植物生长的自然规律，也制作不出盆景精品来。因此，制作树桩盆景的时候，有经验的制作者都会仔细区别树种、判别树形，反复观察、琢磨研究，并在

their design according to actual shape of their source material. Each stump type has its own properties and shapes, thus the processing should go with its natural growth and keep simple. Transforming and shaping the stump against its nature by force should be avoided. If the creators have forced to shape the stump into large-tree style, slanting style or curved style, they will not only violate the law of nature but also fail to create a bonsai masterpiece. Thus when creating a stump bonsai, experienced creators will carefully identify the variety

- 枫树盆景

三角枫属于槭树科，落叶乔木；树皮暗灰色，片状剥落。叶倒卵状三角形、三角形或椭圆形，长6—10厘米，宽3—5厘米，叶片呈三叉状，顶端短渐尖，全缘或略有浅齿，表面深绿色。发叶后开花，颜色为黄绿色，花期4—5月；果实呈棕黄色，果期9—10月。

Maple Bonsai

Trident Maple belongs to the family Aceraceae. It is a deciduous tree with dark grey bark that exfoliates. The leaves are obovate-triangle, triangle or elliptical, 6 cm-10 cm long and 3 cm-5 cm wide and usually with three lobes. Its foliage has short and acuminate apex, with full or serrated margins and dark green surfaces. After the leaves are formed, the yellow-green flowers bloom and the blooming period is from April to May. The fruits are brownish yellow and the fruit period is from September to October.

- **真柏盆景**

  真柏属柏科,圆柏属。常绿灌木,枝干常屈曲匍匐,小枝上升作密丛状。刺形叶细短,通常交互对生或三叶轮生,长3—6毫米,紧密排列,微斜展。球果圆形,带蓝绿色。

  Juniperus chinensis Bonsai

  Juniperus chinensis belongs to the genus *Juniperus*, family Cupressaceae. It is an evergreen shrub with its crooked and creeping trunk and upright branches that are like dense bush. It has thorn-like leaves which are slender and short, usually decussate or three-leaf whorled. The leaves of *Juniperus chinensis* are 3 mm-6 mm long, closely arrayed and slightly extended in an oblique way. The corns are round and bluish green.

画纸上画出蓝图来。即使不落实在纸上,也要在脑海中描绘出大体的轮廓,形成一个较完整的构思,然后才会按构思进行修剪。

of the bonsai tree and analyze its shape through repeated observation, due consideration and study, and then put down the design on paper. Even if they don't draw out the design, they have to draw a rough shape in mind to form a rather complete conception, so that they can start pruning and trimming based on the conception.

蟠扎

蟠扎是盆景造型的一种方法。根据使用材质和方法的不同,蟠扎分为棕丝蟠扎和金属丝蟠扎。

棕丝是棕榈树叶鞘的纤维。用棕丝编结成的绳子对树木的枝干进行造型,就是棕丝蟠扎。棕丝坚韧而具弹性,在蟠扎树木枝干时不易留下痕迹,对树木伤害较小。

棕丝蟠扎的方法是:用棕绳的中段缚住需要弯曲的枝的下端(或打个套结),将两头棕绳相互交叉绞几下,呈麻花状,放在需要弯曲的枝干的上端,打一活结,将枝干慢慢弯曲至所需弧度,再收紧棕绳将活结打成死结,这样就完成了一个弯曲。

金属丝蟠扎常用的金属丝有铁丝、铜丝、铅丝。铁丝在使用前应在火上烧一下,使其硬度变低,容易弯曲。

金属丝蟠扎的方法是:固定好起点,然后用手把金属丝和枝干捏紧,使金属丝和枝干成45°角,拉紧金属丝紧贴枝干的树皮徐徐缠绕。缠绕的密度要均匀,同时还要注意金属丝缠绕的方向。如欲使枝干向右弯曲,金属丝应顺时针方向缠绕;如欲使枝干向左弯曲,金属丝则逆时针方向缠绕。

Wiring

Wiring is a method to style bonsai. According to the different materials and techniques employed, wiring is categorized into two types, namely wiring with palm fiber and wiring with

- **蟠扎的盆景**
  Wired Bonsai

metal wire.

Palm fibers are obtained from palm leaf sheaths. Weaving palm fibers into ropes to style bonsai trunks and branches is called wiring with palm fibers. Palm fiber is characterized by its tenacious strength and resilience. It leaves almost no trace on the bonsai tree after wiring, which makes wiring less harmful to the tree. Wiring with palm fiber follows the following steps: Use the middle of the rope to bind the lower end of the target branch (or make a noose). Twist the rope ends a few times and make a slipknot with them at the upper end of the target branch. Slowly bend the branch into the desired curve, tighten the rope, and then make the slipknot into a dead knot. In this way an artificial curve is finished.

Wiring with metal wires usually employs iron, copper and aluminum bonsai wires. Iron bonsai wire should be heated on fire before wiring to reduce its hardness and make it easier to bend.

The wiring is carried out in following steps: fix one end of the wire. Press the wire against the target branch. Pull the metal wire to wrap the target at a 45 degree branch gently yet tightly against the bark. The wires should be evenly spaced. The direction of the wrap of the wire should also be noted. For example, if one needs to bend a branch to the right, the wrap of the metal wire must be twined clockwise. Nevertheless, if one needs to bend the target branch to the left, the first wrap must be wound anticlockwise.

树桩盆景的养护

一般来说，树桩盆景要放在通风透光，有一定湿度的空间内。对于一些非耐寒性的树桩盆景，在冬天还要搬入温室进行维护。

盆景中的树木如果任其自然生长，不加抑制，势必会影响树姿的造型。所以，要及时对其进行修剪。常采取的修剪方法有：摘心，

Maintenance of Tree Stump Bonsai

In general, tree stump bonsai should be kept in light, well-ventilated and humid places. Some types of tree stump bonsai are not cold tolerant, thus they need to be moved into greenhouses for maintenance and management in winter time.

If left in the state of natural growth without control, the bonsai trees will definitely lose its styled shape. Therefore,

即将其枝梢的嫩头摘去。摘芽，即除去枝干上生长出的不定芽。摘叶，即除去枝干上的部分新叶。修枝，即修剪枝条。修根，即剪掉密集的根系，去掉老根。

此外，给盆景及时补充水分和养分也是必要的。一般说来，夏季或干旱时，早晚应各浇一次水；春秋季节，每天或隔日需浇一次水；春天树桩萌动，还要根据情况早晚浇一次水。至于养分，由于树

timely pruning is necessary. Common pruning methods include: pinching, namely cutting off the growing branch tips; bud pruning, namely pinching adventitious buds; leaf pruning, namely cutting off some newly-grown leaves; branch trimming, namely pruning branches off bonsai trees; and root trimming, namely cutting off the dense root masses and old roots.

In addition, timely water and fertilize for bonsai is also necessary. Generally speaking, in summer or dry time, the grower will water the bonsai twice a day in the morning and evening, while in spring and autumn, one will water the bonsai once every day or every other day. In spring when the stump begins to bud, the

● 树桩盆景
Tree Stump Bonsai

桩盆景比较小，施肥不能太多、太频繁，不仅要掌握施肥的含量和种类，还要把握施肥的季节。

grower will water the bonsai in the morning and evening as well, based on the actual situation of the bonsai. In terms of nutrient supply, considering the relatively small size of tree stump bonsai, the nutrient supply should be neither too much nor too frequent. The grower should know well not only the content and type of fertilizer but the right season to fertilize bonsai.

- 海棠盆景
  Chinese Flowering Crabapple (*Malus spectabilis*) Bonsai

中国榕树盆景之乡——沙西镇

沙西镇位于福建漳浦县西南部,这里气候温和,适宜榕树生长,而且还有着丰富的砂资源,是培育榕树必要的营养土配制材料。2000年6月沙西镇被国家林业局、中国花卉协会授予"中国榕树盆景之乡"的美誉。

沙西榕树盆景具有姿态优美、寿命长、抗病性强、易造型、块根丰满、观感好等特点。主要品种有:块根型榕树盆景、气根型榕树盆景和树桩型榕树盆景。

Hometown of Chinese Banyan Bonsai — Shaxi Town

Shaxi Town, located in the Southwest of Zhangpu County, Fujian Province, has a mild climate that is ideal for banyan trees. It is also rich in sand resource, the source material to make necessary nutrient soil for banyan trees. In June 2000, Shaxi Town was given the title of "Hometown of Chinese Banyan Bonsai" by the State Forestry Administration and China Flower Association.

Chinese Banyan Bonsai produced in Shaxi Town is characterized by its graceful style, a long life span, good disease resistance, easiness to style, well-developed root tuber, good view and so forth. The main categories are: tuberous root Banyan bonsai, aerial root Banyan bonsai and stump Banyan bonsai.

- 榕树盆景

Chinese Banyan (*Ficus microcarpa*) Bonsai

> ## 山水盆景

　　山水盆景，又称"山石盆景"，是以各种山石为主题材料，以大自然中的山水景象为范本，经过精选、切截、雕凿、拼接等一系列的技术加工，将其放到浅口盆

> ## Landscape Bonsai

Landscape bonsai, also called rock bonsai, is created with all types of mountain rocks as its main material and modeled after natural landscapes. Rocks are processed through a series of processing techniques including careful

● 水旱型山水盆景
Water-and-land Bonsai

中，展现出悬崖绝壁、险峰丘壑、翠峦碧涧等各种山水景象的盆景。

山水盆景的类型主要有三种：水盆型、旱盆型和水旱型。其中水盆型山水盆景，只是把山石放到浅口水盆中，盆中只盛水不盛土。旱盆型山水盆景，在浅盆中只放山石而不放水，只需将植物及配件点缀在山石上即可。水旱型山水盆景，浅盆中一部分是土壤、山石、树木，而另一部分则是水。

selection, cutting, chiseling, splicing before being arranged in a shallow basin. This type of bonsai exhibits various landscapes including steep cliffs and precipices, sharp peaks and gullies, green mountains and clear steams and so forth.

There are three main types of landscape bonsai, namely water-filled basin, rock-filled basin and water-and-land basin. Water-filled basin arranges rocks in a shallow basin only filled with water, while the rock-filled basin contains no water but only rocks adorned with plants and decorations. Water-and-land bonsai is partly created with soil, rocks, and trees, with the left part filled with water.

- 旱盆型山水盆景
  Rock-filled Basin

山水盆景的制作

第一，进行艺术构思，确定主题。制作山水盆景时，制作者通常会运用"移天缩地、以小见大"的艺术手法造型和布局。在选择石料制作盆景时，根据石料的特点确定主题。如：修直挺拔、呈悬崖峭壁的，可以用来制作险峰；呈扁长形状的，可以用来表现连绵不断的山峦；皱、瘦、透、漏皆备的，不但

Creation of Landscape Bonsai

The first step is having an artistic conception to determine a theme. To style and arrange one's bonsai, the creator usually employs the artistic technique called "creating a miniature world to imagine the big from the small". When using rocks to create bonsai, the creator determines the bonsai theme according to the characteristics of the selected rocks. For instance, a slender rock with cliff-like sides can be styled as precarious peaks; a flat and rather long rock can be styled to resemble rolling hills; and a rugged, thin, porous and permeable rock not only can be used to resemble a layer of cloud,

- 太湖石

太湖石是制作山水盆景的佳品，具有皱、瘦、透、漏的特点。"皱"指石头表面要有纹理，层叠交错。"瘦"指石形瘦健美，体态窈窕。"透"指石块里要有大小孔道，能够互相沟通。"漏"指石料要有孔隙，能够通气排水。

Taihu Stone

Taihu Stone is ideal for creating landscape bonsai as it is rugged, thin, porous and permeable. A rugged stone refers to a stone with staggered and stepped grain; a thin stone has a vigorous and graceful shape; a porous stone has big and small pore channels, intricately connected; and a stone is permeable because the pores inside allow liquid or air to pass through.

盆景中常用石料

盆景制作中经常使用的石料可以分为两类，一类是质地较柔软的松石，一类是质地坚硬的硬石。松石质松，内有毛细孔隙，易吸水长苔，也易雕琢加工；硬石质坚，不易吸水长苔和点种植物，加工也较困难，但天然纹理较美，形态较奇特，可制成峻峭挺拔的山峰。

常用的石料有：沙积石（砂）、浮水石、芦管石、海母石、钟乳石、英石、太湖石、龟纹石、斧劈石等。

Stones Often Used in Bonsai Creation

Stones often employed to create bonsai are divided into two types. One type is loose stones with relatively soft texture. The other type is hard stones with solid texture. Loose stones, soft and porous, are easy to absorb water and grow moss, which makes chiseling and processing easy as well, while hard stones rarely absorb water and grow moss due to their solid texture. They are also not easy to be dibbled with plants or to be processed. However, hard stones, with beautiful natural grains and unique shapes, can be styled into high and steep peaks.

Stones often used in bonsai creation include sedimentary sandstone, pumice, reed tube stone, corallite, stalactite, Ying stone, Taihu stone, moiré stone, axe chopped stone and so forth.

• 浮水石

浮水石是由火山喷出的岩浆凝固形成的多孔状石块，质轻而松，可以浮在水面之上。这种石料吸水性很强，适宜附植草木，主要产于吉林、黑龙江等地。

Pumice

Pumice, a type of porous stone, forms when magma ejected from a volcano cools down. Light and soft, it floats on water. Due to its good capability to absorb water, pumice is suitable for growing vegetation. Pumice is mainly produced in provinces such as Jilin, Heilongjiang, etc.

- **沙积石**（图片提供：FOTOE）

  沙积石又称"吸水石"，呈灰褐色或棕色，质地疏松，吸水性强，石上可栽植野草、藓苔，主要产于江苏、安徽、浙江、湖北等地。

  ### Sedimentary Sandstone

  Sedimentary sandstone, also called water-absorbing stone, is taupe or brown. Due to its loose and porous texture as well as a good capacity to absorb water, sedimentary sandstone can be planted with weeds and moss. It is mainly produced in Jiangsu, Anhui, Zhejiang, Hubei, etc.

- **钟乳石**

钟乳石是碳酸盐岩地区洞穴内形成的不同形态的碳酸钙沉淀物的总称。其质地坚硬，不吸水，一般用于塑造山岭，主要产于云南、广西等地。

Stalactite

The products of calcium carbonate deposition formed in the caves in carbonate rock areas, are called stalactites that have a solid texture. Stalactite doesn't absorb water and it is generally employed to resemble mountains. Stalactite is mainly produced in Yunnan, Guangxi, etc.

- **海母石**

海母石又称"珊瑚石"，是海洋贝壳类生物遗体积聚而成的化石。颜色一般呈白色，质地疏松。此种石料因含盐分较多，需用清水多次漂洗，才能附生植物。主要产于福建沿海一带。

Corallite

Corallite, also called coral stone, is the fossil formed from accumulated remains of shellfish. It is generally in white with a loose texture. Due to its salt content, corallite cannot be planted with vegetation until it has been rinsed out with clean water. Corallite is mainly produced in Fujian coastal area.

- 英石

英石又称"英德石",是石灰石长期受自然风化侵蚀而成,颜色呈白色,石质坚硬,纹理细腻,不吸水,多用于塑造崇山峻岭,主要产于广东省英德市。

Ying Stone

Ying stone, also called Yingde stone, is formed from limestone eroded over time by wind. It is white and solid with smooth grain. It doesn't absorb water, and is mostly used to resemble high and steep mountains. Ying stone is mainly produced in Yingde City, Guangdong Province.

- 太湖石

太湖石是石灰岩遭到长时间侵蚀而形成的,质地坚硬,体态嶙峋透漏,是园林造景和盆景布置的理想用石。主要产于江苏太湖附近。

Taihu Stone

Taihu stone is the limestone eroded over time by wind and water. It is solid, rugged, porous and permeable, and therefore ideal for garden landscaping and bonsai decoration. Taihu stone is mainly produced in areas near Taihu Lake in Jiangsu Province.

可以制成云峰、洲岛、土山，还可以作为独石欣赏。

第二，选择合适的山石。山水盆景能够将自然景色浓缩到浅盆之中，营造出一种山水画的意境之美，因此，在制作的过程中选择山石就需要一定的技巧。制作山水盆景的石料不仅要具有天然的纹理和色彩，还要具有独特的形态之美。

第三，石料的加工。选取了合适的石料后，制作者就要对其进行必要的加工。加工的时候，要做到自然而无人工痕迹，雕琢要生动流畅、简洁有韵。加工石料的时候，需要用到的工具主要有锤子、凿子、钢锯、刻刀、筛子、小铲和刷子等。同时，还需要一些材料，如水泥、黄沙和颜料等。具体加工的方法有锯截、雕琢、胶接等。

第四，山石的布局。布局是盆景制作的重要步骤，石头的布局可以采用三种方法：独石、子母石和群石。

独石，俗称"孤峰"。就是在盆的左边或右边安放一块较大的石峰，另一侧安放一两块小石作为岛屿，大山、小岛各在其

an islet in a river, or an earth piled hill, but also can be displayed as sole-stone appreciation.

The second step is making a proper choice among one's rocks. Landscape bonsai can exhibit part of the natural scenery in a shallow basin and deliver the beauty of landscape painting, which requires certain skills to select rocks during the creation process. The rocks selected to create landscape bonsai should possess natural grains and colors, with unique beauty in its shape and style.

The third step is processing the selected stones. After selecting the proper stones, the creator needs to process those stones where necessary. The processing is required to be natural, leaving no trace of artificial work. The workmanship needs to be vivid, simple and distinct in an effortless way. The tools needed to process stones mainly include hammer, chisel, steel saw, burin, sieve, small shovel, brush, etc. At the same time, other materials are also needed such as cement, yellow sand, pigment, etc. The processing methods include sawing, chiseling and splicing.

The fourth step is arranging selected stones, which is a step of great importance

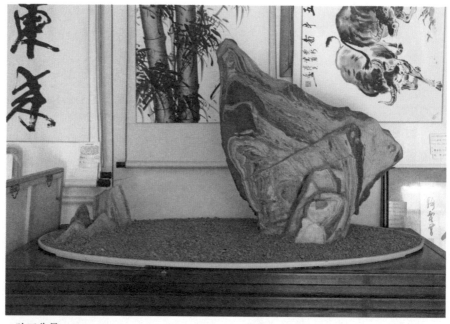

• 独石盆景
Single-peak Bonsai

位,就好像一座山峰矗立在辽阔的水面上一样。这种盆景寓意深远,主题集中。这种石料以自然形成的独石为佳。

子母石,就是在盆里设置一大一小两块石峰,左右对峙。其中要突出母石(主峰),略偏于盆的任意一方,以子石作陪衬。两者高矮不一、大小各异、宾主分明。

群石,由多座较大的石山组

for bonsai creation. In terms of the layout of stones, the creator has three possible choices, namely single-peak, child and mother peaks and multi-peaks.

Single-peak bonsai is also called sole-stone bonsai. To create a sole-stone bonsai, the creator lays a big stone peak in the left or right side of the basin with one or two small stones on the other side to resemble islands. Well arranged, the single peak and small islands resemble a

成，群石中的主山必须摆设在重要的位置上，其体积、高度占绝对的优势。在拼接山石时，要突出主景，宾主有别。

第五，进行山石盆景的配种与配置。当山石布置好以后还要种植一些草木，嵌上一些苔藓和配置一些人物用以点缀。这样，盆景才会

picture of a towering mountain standing by a vast expanse of water. This type of bonsai is considered profound with a highlighted theme. Ideal material to create a single-peak bonsai would be natural sole stones.

Child and mother peaks bonsai keeps two stone peaks in a basin. One stone is bigger and the other is small, facing each other. The creator highlights the mother stone (main peak) by placing it close to the middle of the basin with the smaller stone by its side. The two selected stones vary in both height and size, which makes clear the primary stone from the secondary stone.

Multi-peaks bonsai is composed of several big stones. The main peak, absolutely prominent with regards to its size and height, must be placed at the primary location. When splicing the stones, the creator needs to emphasize the main view and make it easy for bonsai viewers to tell the front from the back.

● 子母石组成的盆景
Child and Mother Peaks Bonsai

变幻无穷，生动活泼，富于诗情画意。配种植物和配置小景时，不但要符合山石和草木的比例，还要注意色彩的和谐，不能任意安放，要有选择、有计划地分布。为了丰富盆景的内容，在山石的缝隙、阴暗面和山脚的地方，可以栽植一些常

The last step is planting trees and adding decorations. After arranging the mountain stones in the basin, the creator needs to plant vegetation, pave some moss and add some figurines of people as decorations. In this way, bonsai provides so many possibilities in a vivid and poetic way. When planting vegetation and adding views to bonsai, one should notice the proportion of stones and vegetation, as well as the harmony between different colors. The locations of plants and decorations should not be chosen randomly. Instead, they should be carefully chosen according to the existing plan. To enrich the content of bonsai, one

- 群石组成的盆景
  Multi-peaks Bonsai

● 山水盆景（图片提供：全景正片）
Landscape Bonsai

绿的小草木、青苔，还可以配置一些必要的舟楫、桥梁、亭榭等。

can plant some evergreen vegetation and moss in stone cracks, in the shades and at the foot of mountains. Boats, bridges and pavilions can also be added where necessary.

山石的加工

在对一盆山水盆景进行选材时，首先要对石材的顶部轮廓线进行观察，引发构思，反复推敲。不论硬石、软石，在轮廓线不明显时，都要对其进行敲削。其次还需要截锯与黏合。石材一般都是天然未经加工的，必须根据景观要求对石材进行锯截、连接和黏合。锯截石材用切割机或钢锯进行，也可以用锤子敲断、理平的方式进行。然后，再用水泥或水泥兑色，或其他黏合剂黏结。

此外，一盆山水盆景要求在石体皴纹上达到大体一致，这样显得景观画面较为统一。在尽量保持纹理自然、线条一致的前提下，有时要对一些纹理皴法不明显或纹理差异较大的石材进行理纹。理纹一般用剔、掏、敲、锯等方式进行。

Processing Stones

When selecting stones for a landscape bonsai, the creator first needs to observe the top contour line of stones to construct the conception with repeated consideration. Both hard and soft stones need to be struck and whittled, and then to be cut and spliced if their contour lines are not clear. Generally, stones are natural and not processed, therefore it is necessary to cut, connect and splice the material to meet what bonsai views require. Stones are cut with cutters or steel saws, or struck to split and flattened by hammers. Then the stones are spliced with cement or colored cement, or other bonding agents.

In addition, stones selected for the same bonsai are required to be roughly unified in terms of shades and textures for a moderately coherent view. On the premise that all stones are selected with their natural grain as unified as possible, sometimes grain combing needs to be conducted to stones with unclear or very different grains, shades or surface textures. Grain combing is commonly conducted through combing, scooping, striking and sawing.

• 山水盆景
Landscape Bonsai

《云林石谱》

《云林石谱》是中国第一部论石专著，由宋代的矿物岩石专家杜绾编著。全书一万多字，涉及名石一百多种。作者详细考察了这些名石的产地，还细数其采取方法、形状、颜色、质地优劣、敲击时发出的声音、坚硬程度、纹理、光泽、晶形、透明度、吸湿性、用途等方面的特点。

Yunlin Stone Collection

Yunlin Stone Collection, China's first monograph on stones, was compiled by Du Wan, an expert on minerals and stones in the Song Dynasty (960-1279). This book has over 10,000 characters and describes over 100 major types of stones. The author has conducted field investigations in producing areas and kept down different characteristics of stones including the quarrying method, shape, color, texture, knocking sound, rigidity, grain, gloss, crystalline form, transparency, moisture absorption, usage, etc.

萤石
Fluorite

山水盆景的养护

山水盆景，泥薄、水浅，植物幼嫩，在保养上需要倍加小心。为了便于植物生长，平时要将山水盆景放在半阴阳、向南通风而又方便观赏的地方。夏季气温高，不能晒得太久，必须遮挡烈日，以免幼小植物旱死。

Maintenance of Landscape Bonsai

The maintenance of landscape bonsai requires intensive care, due to its thin layer of soil, the shallow water in the basin, and the young and fragile vegetation. To help bonsai plants grow, landscape bonsai ought to be kept in well ventilated places facing the south, with indirect sunshine. It needs to be easy

为了保持水的清洁，盆中不但不能断水，而且要经常换水。同时，还要时不时地在山石上淋水，确保山上的植物不缺水，保持青苔和草木的青翠。如果植物生长旺盛，还要进行修剪、整容。

to appreciate as well. In summer time, due to the high temperature, landscape bonsai cannot be left in the sunshine for too long. Instead, it needs to be sheltered from the burning sun in order to prevent young vegetation from being dried up.

To keep water clean in the basin, the grower can never leave the basin to dry. One even needs to change the water rather frequently. At the same time, the grower also needs to mist the stones from time to time, to make sure that bonsai plants have adequate water supply. Therefore, the moss and vegetation will stay fresh and green. When bonsai plants are growing too rapidly, pruning and trimming ought to be carried out.

• 盆景园一角
A Glimpse of the Bonsai Garden

> 水旱盆景

水旱盆景是介于树木盆景与山水盆景之间的另一类盆景形式，是以树木、山石、人物、水、土为材料，采取山石隔开的方式制作出来的各式各样的树景。成形后的盆景，不仅有水面、旱地，还有树木和山石，给人以联想。

水旱盆景的制作

首先，选择恰当的盆。一般来说，最好选用浅口的石盆，形状不能太复杂，长方形或椭圆形最好；色彩不能太缤纷，白色最好。

其次，选择合适的树木材料和石料。一般情况下，要选择小叶的树种，如榔榆、雀梅、六月雪、虎刺、黄杨等。然后，经过一定时期的培育，使之初步成形。孤植时，

> Water-and-land Bonsai

Water-and-land bonsai is a bonsai style sharing some similarities with both tree stump bonsai and landscape bonsai. It employs trees, mountain stones, figurines, water and soil as main materials. It creates different tree scenes that are separated by mountain stones. A styled bonsai not only exhibits water and land, but also displays trees and rocks in its basin, leaving the viewer space of imagination.

Creation Water-and-land Bonsai

The first step is choosing the proper basin. Generally speaking, it is better to choose a stone shallow basin. The shape of the basin should not be too complex: square or oval would be ideal. The color of the basin also cannot be too complex: white is the best.

The second step is to choose proper

- 水旱盆景（图片提供：全景正片）
  Water-and-land Bonsai

树木形态要力求完整；合栽时，要注意搭配效果。

　　用作水旱盆景的石料，最好是硬质的，如英石、龟纹石、卵石等。也可以采用经过适当雕凿的软石，如砂积石、芦管石等。一般说来，石料形状要圆浑，不能多棱角；要平常，不能太奇特。此外，同一盆景中的石料，在纹理和形状

trees and stones. In general, the grower will choose small leaf trees, such as Chinese elm, *Sageretia theezans*, Junesnow, *Damnacanthus indicus*, Chinese boxwood, etc. The plant will develop a primary shape after being cultivated for a certain period. Solitary planting should keep the integrity of the plant shape, while group planting should focus on the overall effect.

上要力求统一。

再次，加工树木材料和石料。树木加工时，要着重考虑配置效果，要根据盆景总体造型的需要进行取舍。对石料加工，主要是将多余的部分切除。有时为了达到盆景的有效统一，有些石料在切截后还需做适当的雕凿。

最后，布局摆放。石料与树木加工完成之后，就可以进行布局摆放了。将各种材料配置在一起，放

The most favorable choice of stones for water-and-land bonsai is hard stone, such as Ying stone, moiré stone, pebble, etc. Soft stone after being properly chiseled can also be adopted, such as sedimentary sandstone and reed tube stone. Generally speaking, the selected stones ought to be round and smooth stones rather than pointed stones; ordinary stones are preferred while odd ones need to be avoided. In addition, the grain and shape of the selected stones in

● 水旱盆景
Water-and-land Bonsai

● 水旱盆景
Water-and-land Bonsai

进一些土壤，通过反复审度调整，取得理想的效果。布局初步定下之后，就可以将树木和石料依次放入盆中。栽种树木时要注意位置、角度和方向。布置点石时，要注意高低、疏密等变化及与其他景物的关系。最终的效果要使树木与山石形成一体，就像天然生成的一样。

有时为了突出主题，增添生活

the same basin should be unified.

The third step is processing trees and stones. With regards to processing trees, the grower needs to focus on the effect of different combinations. Decisions on whether to adopt or discard a view depend on the whole style of bonsai. Processing stones mainly is to cut off unnecessary parts. Sometimes, stones need to be chiseled after cutting for an overall unified style.

The fourth step is arranging bonsai layout. After processing trees and stones, the grower will start arranging bonsai layout, including arranging the selected materials in groups, putting some soil in the basin, and then making adjustment after repeated observation. After the grower has made the primary decision on the layout, one can put trees and stones into the basin in proper order. Location, angle and direction should be noted when planting trees. Changes in height and density, as well as relation with other elements should be noted when arranging stones. In this way the final effect of combining trees and stones will reach a natural harmony.

In some cases, to highlight the theme and add some flavor of life to bonsai art, the grower often employ some bonsai

气氛，水旱盆景中经常会使用一些房屋、亭台、人物、动物、小桥、小船等配件作为点缀。

decorations such as miniature buildings, pavilions, figurines of people and animals, bridges, boats, etc.

盆景中的配件

在盆景的制作过程中，为了丰富盆景景观，深化盆景意境，使盆景的整体效果更加完美，制作者往往会在盆景中加上一些配件。常见的配件有：动物造型，如鱼、鸟、牛、马、鹤；人物造型，如农夫、樵夫、牧童、仕女；建筑造型，如寺庙、亭台、楼阁、塔桥。

点缀配件时应注意，首先，需要根据不同的造型选择不同的摆放位置。一般亭台要放置在山腰上，塔要放置于次峰之上。其次，配件的数量不宜过多，否则会杂

- **人物点缀的盆景**（图片提供：全景正片）
  Bonsai Decorated with Figurines of People

乱无章。一般用一两件点缀盆景就可以了。第三，还要注意配件与盆景的协调。配件在盆景中起到的是画龙点睛的作用，所以配件的色泽与盆景的色泽要有所区别。

Bonsai Decorations

In the process of bonsai creation, the grower often adds some decorations in one's work to enrich its views, deepen the artistic conception and perfect its overall style. Common decorations include: figurines of animals, such as fish, bird, cattle, horse and crane; figurines of people, such as farmer, woodman, shepherd boy and beautiful lady; miniature buildings, such as temple, pavilion, storied building, tower and bridge.

The following should be noted when adding decorations. Firstly, choose the locations of decoration according to different styles. In general, a pavilion is decorated half way up the mountain, while a tower is placed on the secondary peak. Secondly, the number of the decorations should be controlled. If not, the bonsai will be disorganized. Generally one or two decorations for one bonsai will be enough. Thirdly, the harmony between decorations and bonsai should be noted. Decorations add the finishing touch in styling bonsai; therefore, the colors of decorations should be different from the color of bonsai.

水旱盆景的养护

水旱盆景平时要放在通风透光处。夏季要防止暴晒，冬季需移至室内光照充足处，不能受冻。

水旱盆景因用盆很浅，特别需要及时补充水分。为了防止浇水时盆土被冲掉，人们通常都会用细眼喷壶浇水。气候干燥时，为了保持苔藓生长良好，还应用喷雾器在盆面及树木、石头上喷雾。

Maintenance of Water-and-land Bonsai

This type of bonsai should be kept in well-ventilated and light places. In summer it should be kept out of direct sun, while in winter it should be moved to places with abundant sunlight and protected from cold temperatures.

As bonsai is kept in a shallow basin, a timely moisture supply is highly necessary. Growers will water their

水旱盆景中的土壤通常都较少，养分有限，为了使树木生长健壮，通常都会为其施肥。施肥时，可用稀释后的有机肥水对土面进行喷洒，或将颗粒状有机复合肥埋入盆土中。

水旱盆景中的树木一般3—5年换一次土，多在春、秋季进行。换土时，通常都会先将点石及配件取

bonsai with fine watering cans in case that the soil in the basin will be washed away. In dry times, to keep the moss in a good growing state, growers should mist the basin, trees and stones with sprayers.

Nutrients are capped due to the limited amount of soil kept in the basin. Therefore, growers usually apply fertilizer to bonsai trees to help them grow strong. With regards to applying

● 水旱盆景——国魂（图片提供:FOTOE）
Water-and-land Bonsai — National Spirit

• 水旱盆景（图片提供：FOTOE）
Water-and-land Bonsai

下，将树木连土小心地拔出来；然后用竹签剔除约二分之一的旧土，剪去死根及部分老根，换上新土；接着再按照原来的位置栽进盆中，安放配件，铺上苔藓，这样就完成了。

fertilizer, the grower either sprays water with diluted organic fertilizer onto the soil surface, or buries granulated organic compound fertilizer in the soil.

The soil kept in the basin is changed every three to five years, mostly in spring or autumn. When changing the soil, the grower will firstly take decorated stones and other decorations away, carefully take the trees out with soil. Then one will comb about half of the old soil off with a bamboo skewer, cut off dead roots and some old roots, and put some new soil into the basin. At last, the grower will plant the trees back into the basin at their original places, put the decorations back and pave stones with moss.

黄果树盆景园

　　黄果树盆景园地处贵州西部的低洼地带，海拔较低，终年无霜，适宜亚热带植物生长。盆景园里到处都长满了仙人掌、黄兰、白玉兰等多种植物及其他喜温灌木乔木。除此之外，南方植物园里常见的栾树、米兰等也可以在这里看到。园内盆景风情万种、花红叶翠、交相辉映、布局天然，园内展示着各种姿态各异的珍奇盆景3000多盆，各个争奇斗艳、竞吐芬芳。

Huangguoshu Bonsai Garden

The Huangguoshu Bonsai Garden is located in a low-lying area in western Guizhou Province, which is also a low altitude place. It is frost-free all the year round, and therefore suitable for subtropical plants to grow. Many species of plant including cactus, champak, magnolia, and some other thermophilous bush and arbor tree species all grow here. In addition, goldenrain tree and Chinese perfume plant that are commonly seen in botanical gardens in South China can also be seen here. The bonsai in the Huangguoshu Bonsai Garden are all fascinating and charming with their rich colors and natural-like beauty. There are over 3,000 rare bonsai species displayed in this garden in various styles with pleasant fragrance.

- 贵州黄果树盆景园一角（图片提供：FOTOE）
A Glimpse of the Huangguoshu Bonsai Garden in Guizhou Province

> 花草盆景

花草盆景是以花草或木本的花卉为主要材料，经过一定的修饰加工，适当配置一些山石和点缀配件，表现花草的景观。成型后的花草盆景能够以小见大，将自然界优

> Flower Bonsai

Flower bonsai employs flowers or woody plants as main material, which are processed and then exhibited with mountain stones and other decorations. Styled flower bonsai will fully deliver the beauty of flowers and plants in nature through the small part of nature exhibited in the pot.

Selecting material for creating flower bonsai is not only different from creating tree stump bonsai, but also different from planting potted flowers. Possible choices include woody flowers such as Chinese flowering crabapple, Chinese rose, azalea, camellia, winter jasmine and so forth, as

• 盛开的花朵
Flowers in Full Bloom

美的花草景色充分展现出来。

花草盆景的选材既不同于以观赏树姿为主的树桩盆景，也有别于一般的盆花，既可以选用木本花卉，如海棠、月季、杜鹃、山茶、迎春等，也可以选用草本植物，如兰花、菊花、水仙等。

花草盆景的制作

制作花草盆景，可以根据自己的构思将花草在盆中设计出理想的布局。花草盆景注重的是整体的韵味。制作者往往都非常重视其中的意境，通过巧妙的修饰，配置好山石和配件，就能将自然优美的花草景色表现出来。

用来制作花草盆景的植物，一般都要姿态优雅，叶形美观，色彩亮丽。常用的植物有：兰花、菊花、文竹、翠竹、迎春、栀子、水仙等。花草位置确定后，还可以适当点缀一些山石来完善构图，丰富意境。

花草盆景中点缀的山石，通常都会充分利用它的自然形态。常用的石种有英德石、石笋石、芦管石、钟乳石、龟纹石、昆山石、宣

well as herbaceous plants such as orchid, chrysanthemum, daffodil and so forth.

Creation of Flower Bonsai

To create flower bonsai, one can design an ideal layout of flowers based on personal conception. Flower bonsai focuses on the overall flavor. Creators usually pay great attention to one's conception and therefore they are capable to exhibit the beautiful scenery of flowers in nature, through smart embellishing and well arranging mountain stones and other decorations.

The selected plants for creating flower bonsai are generally elegant and colorful with graceful leaf shape. Favored choices include orchid, chrysanthemum, *Asparagus setaceus*, *Sasa pygmaea*, winter jasmine, cape jasmine, daffodil, etc. After determining the places of flowers, the grower can also decorate flower bonsai with mountain stones properly, to improve the composition of the whole picture of one's bonsai as well as to enrich its conception.

When selecting mountain stones as decorations for flower bonsai, the creator will usually fully utilize their natural shapes. The commonly used

石、灵璧石等。制作花草盆景，要表现出盆景的雅致和韵味。如果植物是大花大叶的种类，比较笨拙，那么石头就要选用一些精巧、奇特的假山型；如果植物是兰花之类的小叶型，品质淡雅，形体精巧，就要选一些外形古朴的山石。

盆中景物布置就绪后，要在土面铺上一层细细的青苔，使景色更显清新自然。此外，一个精美的花盆也是衬托各种花草的优美形态和瑰丽色彩的必不可少的要素。

stones include Yingde stone, stalagmite, reed tube stone, stalactite, moiré stone, Kunshan stone, Xuan stone, Lingbi stone, etc. Flower bonsai are created to deliver the taste and flavor of bonsai art. If bonsai plants are species with big heavy flowers and leaves, it is better to choose exquisite, odd and rockery stones as decorations; while if the plant species are small-leaf plants, for instance, orchid, characterized by simple and exquisite elegance, then plain and unsophisticated mountain stones would be ideal for decoration.

● 花草盆景
（图片提供：全景正片）
Flower Bonsai

京都第一盆景园

　　京都第一盆景园，位于北京通州区半截河村。这个盆景园占地面积18亩（约1.2公顷），是北京单体盆景园中面积最大的。园中收藏有1800多盆盆景，囊括了扬派、川派、苏派、通派、海派等八大派系的盆景。其中，珍贵的桂花盆景大约有500盆、15个品种，最大的桂花盆景的桩龄达110年、盆龄达30年。此外，这里的精品梅花盆景有200多盆、12个品种，最大的盆龄有100余年。

Beijing No.1 Bonsai Garden

The Beijing No.1 Bonsai Garden is located in Banjiehe Village, Tongzhou District, Beijing. Covering a total area of 18 *Mu* (approx. 1.2 hectares), it is the largest single garden for bonsai display in Beijing. There are over 1,800 bonsai displayed in this garden, and all eight major genres of Chinese bonsai are collected including Yangzhou style bonsai, Sichuan style bonsai, Suzhou style bonsai, Nantong style bonsai, Shanghai style bonsai, etc. Among these displayed bonsai, there are about 500 rare sweet osmanthus bonsai of 15 categories. The oldest sweet osmanthus bonsai has a plant age of 100 years and a bonsai age of 30 years. Additionally, the garden keeps more than 200 Chinese plum bonsai of 12 categories, among which the oldest has a bonsai age of more than 100 years.

● 大型室外盆景
Large Outdoor Bonsai

花草盆景的象征意义

花草盆景的主体是花草、花卉，不同的花草盆景有着不同的象征意义。例如：天堂鸟是自由、幸福、快乐、吉祥的象征。如果家有老人做寿，摆放一盆天堂鸟盆景以祝老人似仙鹤般长寿。

石竹代表的是谦虚、多愁善感。其中，单瓣品种被喻为"花中林黛玉"，重瓣品种表现出热情

After all elements have been arranged in the pot, the surface needs to be paved with a fine layer of moss to convey a more fresh and natural view. What is more, an exquisite flowerpot is also essential to give full play to the beautiful styles and glorious colors of all flowers.

Symbolic Meaning of Flower Bonsai

With flowers and plants as the main body, different flower bonsai have different symbolic meanings. For example, bird

• 花草盆景
Flower Bonsai

● 文心兰盆景 （图片提供：FOTOE）
Oncidium Bonsai

洒脱。长寿花叶子肥厚，花繁色艳，一年四季都会开花，也代表了长寿。石蒜代表了优美和纯洁。鸡冠花，在风雪的洗礼下，也不减花姿、不褪花色，寓意"永不褪色的恋情"或"不变的爱"。南天竹，茎杆光滑，红叶满枝，红果累累，

of paradise flower (*Strelitzia reginae* Aiton), shaped like a crane, is the symbol of freedom, well-being, happiness and good luck. Having a bird of paradise flower bonsai, when celebrating the birthday of the elder member in the family, symbolizes long life like the long-lived cranes.

Chinese pink symbolizes modesty and sentimentality. Single-lobe Chinese pink is likened to Lin Daiyu, a tragic figure in *Dream of the Red Chamber*; while double flower species symbolizes an enthusiastic and carefree personality. Jonquil, with fleshy leaves and colorful flourishing flowers that bloom all the year round, stands for long life. Red spider lily represents gracefulness and purity. Cockscomb, even in a wind-driven snow, can still keep its charm and color. Therefore, the message carried by cockscomb is "love never fades" or "love never changes". Heavenly bamboo has smooth stalks, red leaves and fruits hanging heavily. It never withers away

● 水仙盆景（图片提供：全景正片）
Daffodil (*Narcissus*) Bonsai

经久不凋，象征长寿。富贵竹，淡雅、清秀，象征了吉祥富贵。龟背竹，叶形奇特，青翠可爱，寓意健康长寿。

and therefore symbolizes long life. Lucky Bamboo is simple, elegant and delicate, thus it represents good luck, wealth and power. Ceriman is freshly green and lovely with a very unique leaf shape. It carries the message of health and longevity.

可以净化空气的花草盆景

在花草盆景中，有很多能够起到净化空气的作用，具有代表性的有这样一些品种：

白掌盆景。白掌是过滤室内废气的"专家"，可以过滤掉空气中的苯、三氯乙烯、甲醛、氨气和丙酮等。

吊兰盆景。吊兰的枝叶不仅可以有效地吸收窗帘释放出的甲醛，还能够充分净化室内的空气，所以经常被人们放置在浴室、窗台或者搁架上。

散尾葵盆景。散尾葵是最天然的"增湿器"，经常给散尾葵盆景喷水，不仅可以使其保持葱绿，还能清洁叶面的气孔，从而达到净化空气的目的。

绿宝石盆景。绿宝石微张的叶子可以吸收有害物质，并将之转化为对人身体无害的营养物质，适宜种在吊盆中。

常春藤盆景。常春藤能有效吸收吸烟产生的烟雾，从而有效抵制尼古丁中的致癌物质。常春藤长有很多长长的枝叶，只要将枝叶巧妙地放置就可以出现一盆别样的盆景。

黄金葛盆景。黄金葛能抑制吸烟产生的烟雾，通过类似光合作用的过程，可以把织物、墙面和烟雾中释放的有毒物质分解为植物自有的物质。黄金葛长有漂亮的心形叶子，具有独特的装饰性，可以用来装饰居室。

袖珍椰子盆景。椰子是生物中的"高效空气净化器"。它能够同时净化掉空气中的苯、三氯乙烯和甲醛，适合摆放在新装修好的居室中，有效改善室内空气质量。

Flower Bonsai That Can Help Clean Indoor Air

There are many types of flower bonsai that can help clean indoor air. The following are typical ones:

Spathiphyllum kochii bonsai: *Spathiphyllum kochii* is the "expert" to curb waste gases indoor. It can scrubs benzene, trichloroethylene, formaldehyde, ammonia, acetone, etc.

Spider plant bonsai: The branches and leaves of spider plant can effectively absorb formaldehyde released by curtains and thoroughly clean the indoor air. As a result, spider plant bonsai are usually placed in bathrooms, on window-sills or shelves.

Bamboo palm bonsai: Bamboo palm is the best natural humidifier. Spraying water on Bamboo palm bonsai will not only keep it fresh and green, but also clean the stomas in its leaves, which will eventually help to clean the indoor air.

Black bean bonsai: Its slightly open leaves can absorb harmful substances, and then turn

● 常春藤盆景（图片提供：FOTOE）
Common Ivy (*Hedera helix*) Bonsai

those substances into nutrients that have no harm to human body. It's better to raise black bean as hanging plant.

Common ivy bonsai: Common ivy can effectively absorb cigarette smoke and therefore help to curb carcinogenic substances in nicotine. It has many long branches and if they are arranged smartly, a unique bonsai can be created.

Golden pothos Bonsai: Golden phthos can curb cigarette smoke. Through a photosynthesis-like process, the harmful substances released from fabrics, walls and smoke are resolved into substances that compose the plant. With its beautiful heart-shape leaves, golden pothos is considered rather unique as a decoration. It can be employed to decorate living rooms.

Mini coconut bonsai: Coconut is considered as super air cleaner among all living things. It can scrub benzene, trichloroethylene and formaldehyde at the same time, ideal for newly furnished rooms to effectively improve indoor air quality.

- 散尾葵盆景（图片提供：FOTOE）
  Bamboo Palm (*Chrysalidocarpus lutescens*) Bonsai

> 微型盆景

微型盆景是以花草为主，同时用山石等小件作为点缀，配置而成。制作微型盆景时，经常使用的花草有：文竹、虎耳草、吊兰、兰花、万年青、水仙、菊花、芭蕉、芦苇等。一般来说，微型树木盆景的高度在10厘米以下，盆长通常都不超过10厘米。

微型盆景看重的是形态小巧，造型玲珑别致，更注重整体艺术美的内涵。每一小组盆景的各种摆件，都要表现出各异的形态，注重整体布局。这种盆景聚散有致，主题突出，具有一定的审美情趣，表现了微型盆景艺术的生命力。

> Mini Bonsai

The main body of mini bonsai is flowers, adorned with small decorations, i.e. mountain stones, and therefore this type of bonsai is called miniature bonsai. When creating this type of bonsai, the creator usually utilize *Asparagus setaceus*, *Saxifraga stolonifera*, spider plant, orchid, dieffenbachia, daffodil, chrysanthemum, *Musa acuminata*, reed and so forth. In general, a mini stump bonsai is shorter than 10 cm, with its pot length less than 10 cm.

Mini bonsai focuses on the small and exquisite shape as well as a unique and delicate style. It greatly emphasizes the artistic connotation of its overall beauty. This type of bonsai is arranged in an organized and aesthetic way with a highlighted theme, which is the art vitality of mini bonsai.

● 微型盆景（图片提供：FOTOE）
Mini Bonsai

微型盆景的制作

制作微型盆景，首先要选择适宜的植物材料。同时，还要配上一个做工精细、造型优美的紫砂小盆，在盆底钻孔后用作盆器。选定植株后，便可以将其栽植到适宜的盆中。一般来说，在长筒盆中，要栽半悬崖式或全悬崖式植株；如果

Creation of Mini Bonsai

To create a mini bonsai, the first step is to select the proper plants. At the same time, the creator needs to prepare a small yet graceful purple clay pot. One will drill a hole in the bottom of the pot and in this way it serves as a bonsai container. Trees can be planted in the proper pots once they are selected. Generally

● 微型盆景（图片提供：FOTOE）
Mini Bonsai

是在长方形及椭圆形盆中，最好选择直干或斜干式植株。

其次，布置好造型。通常来说，植株造型最适宜在入冬以后至来年的春天植株萌芽前进行。在进行造型前，制作者会依据所栽树木的特点和姿态进行构思设计，确定造型方式。

主干，是微型盆景造型的主要部分，如果确定为直干式，则不需要蟠扎，保持干直挺拔，蓄养侧枝即可；如果是斜干式，只要把主干偏斜栽植，就可以了；如果造型是

speaking, tubular containers are suitable for tree stump bonsai of semi-cliff style or cliff style; while square and oval pots are suitable for upright or slanting style.

Secondly, develop a style for bonsai. Generally speaking, the best time to style bonsai trees is from winter until early spring when trees begin to sprout. Before styling the tree, the creator will compose one's conception and design, as well as determine one's styling methods based on the characteristics and shapes of the planted trees.

Trunk is the main part of styling mini bonsai. If the bonsai is determined to be upright style, wiring is not needed. Keeping the trunk upright and straight and holding side branches are enough; and if it is determined to be slanting style, planting the tree with its trunk inclined to one side is enough; while if it is determined to be cliff-style trunk, styling the trunk by wiring it with aluminum wires is a must.

When processing, the creator will not keep the foliage too dense. In general, one only keeps one or two branches to show its natural beauty. All bonsai creators will make every effort to reach a simple and smooth style. In terms of unorganized branches that affect the

- 微型盆景（图片提供：FOTOE）
  Mini Bonsai

是悬崖式，必须用铅丝缠绕主干将其弯曲成型。

加工时，枝叶不必过繁，一般要留1—2个枝条，为了显示其自然美，制作者都会力求简练、流畅。对于交叉枝、平行枝、对生枝、辐射枝、反向枝等杂乱、有碍造型的枝条，人们通常会及时疏剪或短截，使枝条疏密有序、层次分明、高低适度。

微型的山水盆景，经常会用到浮石、斧劈石、芦管石等石料。用盆时，经常选择汉白玉等加工成长方形或椭圆形浅盆。造型制作时，盆内放置的山石不要太多，但也要有主峰、次峰和配峰的区别，为了衬托山石的高大，可以放置一个比例恰当的小点缀品。

微型盆景的养护

微型盆景因土量少，植株弱小，经不起风吹日晒，要放置在阳光充足、避风、遮阴的场所。微型盆景的盆面没有水槽，浇水时，只要将盆放到贮水容器中浸透几分钟即可。

为了使植株枝叶的长势保持

overall bonsai style, such as crossing branches, parallel branches, converged branches, radiating branches and reversed branches, the creator will trim or cut off some branches, in order to keep the branches evenly spaced with clearly identified layers and a proper height.

Mini landscape bonsai usually utilize pumice, axe chopped stone, reed tube stone, etc. Its basin is generally a shallow square or oval basin made of white marble. When the bonsai is styled, the amount of mountain stones placed in the basin should be controlled with the primary peak, secondary peak and the rest peaks clearly identified. A small proportional decoration can be positioned to show the height and size of the towering peaks.

Maintenance of Mini Bonsai

Due to its fragile and small vegetation as well as the small amount of soil, mini bonsai cannot bear strong wind or burning sunshine. It ought to be placed in light places sheltered away from wind and direct sunlight. When watering mini bonsai with no trough in its pot base, one can simply put its pot in a water container for several minutes until it is saturated.

- 微型盆景
  Mini Bonsai

下去，要及时补给养分，还要在深秋或初春翻盆换土。同时，为了控制盆景的生长，要及时对植株进行摘心、摘叶和剪叶。

Timely nutrient supply is required to help bonsai plants keep growing. This should be carried out in late autumn or early spring, along with soil replacement. To control its growth, one needs to pinch, prune and trim bonsai plants timely.

> 挂壁盆景

挂壁盆景，是将传统的盆景艺术同贝雕、树皮画等工艺美术品制作技艺和形式巧妙地结合在一起，并吸收了国画、书法艺术精华而产生的一种新型的盆景艺术品。由于这种盆景不会占用桌面和地面，形式新颖，美化居室效果极佳。

根据取材和制作方法的不同，可以将挂壁盆景分为挂壁式山水盆景和挂壁式树木盆景。

挂壁式山水盆景的造型方法和一般山水盆景一样，常用大理石板作为背景，如果大理石板上具有天然的抽象的山水纹理更理想。如果近景要种植一些较大的植物，可将容器置于底板背后，在底板隐蔽处打个孔，植物从孔中穿过，根部在后，冠部在前。如果只配植

> Wall-hanging Bonsai

Wall-hanging bonsai smartly combines traditional bonsai art with the workmanship and forms of handicraft art works such as shell carving and bark picture. It is a new type of bonsai art which has absorbed the artistic cream of traditional Chinese painting and calligraphy. It does not occupy desk or floor area and is considered original and ideal as a room decoration.

According to the difference in material selection and processing, wall-hanging bonsai is categorized into two types, namely wall-hanging landscape bonsai and wall-hanging stump bonsai.

Wall-hanging landscape bonsai is styled as ordinary landscape bonsai, with dolomite employed as its backdrop. Dolomite board with landscape grains would be ideal. If the grower needs to plant larger plants for in the close view,

- 挂壁式盆景（图片提供：FOTOE）
  Wall-hanging Bonsai

one can place the container behind the backdrop board. Then the grower needs to choose an inconspicuous place of the board to drill a hole through which trees are planted with roots hidden under the board and crown shown outside. Small vegetation can simply be planted in soil in the cracks of stones with decorations that are placed according to the requirements of the overall conception.

Trees with a slow growth rate, good adaptability, graceful shape and some shade tolerance are usually selected for wall-hanging stump bonsai. There are two types of layout for wall-hanging stump bonsai. One layout is hiding the container behind the planted vegetation. Drill a hole in a proper place of the basin; stick a half container at the back of the basin slightly below the hole and put some compost in the container where the roots of plant will later be kept. In this way, viewers can only see the beautifully shaped and styled plants from the front

微型植物，可以直接植于山石缝隙的泥土中，然后再按意境要求点缀配件即可。

挂壁式树木盆景选用的树种通常都生长缓慢、适应性强、树姿优美，具有一定的耐阴能力。挂壁式树木盆景的布景形式有两种：一种是将盆藏于所栽植株背后。首先在盆器的适当部位凿一个洞，在盆背

沈阳盆景园

　　沈阳盆景园位于辽宁省沈阳市，占地面积为2.6万平方米，是目前中国北方地区最大的盆景展示园。盆景园里山地起伏变化，各种各样的树木、假山、叠石呈现出了多姿多彩的效果。盆景园内设有四个小园：梅园、兰园、竹园和菊园。每个小园的景墙风格各不相同，有的是用自然气息浓厚的天然文化石建成，有的包含了中国传统的木制漏窗设计元素。园中的小路则全部用浙江青石雕刻成了不同流派、形态各异的盆景浮雕。

Shenyang Bonsai Garden

Shenyang Bonsai Garden is located in Shenyang City, Liaoning Province. Covering an area of 26,000 square meters, it is the largest bonsai display garden in Northern China at present. The garden is a hilly area with various trees, rockeries and piled stones kept in the garden. There are four small gardens named after plum blossoms, orchid, bamboo and chrysanthemum in the bonsai garden. The walls of the four small gardens have different styles. Some are built with natural culture stones that carry a flavor of the great nature, and some embrace the design concept of traditional Chinese wooden lattice window. The roads in the bonsai garden are paved with Zhejiang blue-stones, carved in relief into bonsai patterns with various styles of different bonsai schools.

● 树桩盆景
Tree Stump Bonsai

● 树桩盆景
Tree Stump Bonsai

面、洞口略下方粘贴半个容器，盆内放些培养土，树根从洞孔穿过栽入容器中。这样，从正面只能看到优美的树姿造型，看不到容器的裸露部位。另一种是将盆器和树木姿态完全显露出来。首先选用一个已经成型的悬崖式或曲干式小型树木盆景，然后在大小适宜的背景板上选一个适当的位置，按盆器的形状锯一个小洞，正好将盆放入其中。最后，在板的正面最多露出三分之一，在板的背面将盆器固定。

and they cannot see the container. The other one is fully presenting both bonsai plants and the bonsai basin. Firstly select an already styled stump bonsai of cliff or curved trunk style; then cut a hole in the backdrop board. The shape and size of the hole should be trimmed to fit the basin in; the last step is fastening the basin to the board from the back, with at most one third of the basin revealed in the front.

> 异形盆景

异型盆景是将植物种在特殊的器皿里，经过精心养护和造型加工，做成的一种别有情趣的盆景。这种盆景的造型自然，不拘一格。

异型盆景一般体量都不太大，

> Fancy Bonsai

Fancy bonsai keeps plants in special containers. Through elaborate maintenance, styling and processing, fancy bonsai is created with a different appeal. It is styled in a natural way without sticking to a certain pattern and

- **笔筒盆景**（图片提供：全景正片）

笔筒是文房用具之一，古时候，笔筒是搁放毛笔的专用器物。笔筒的材质有木制的、玉石制的、瓷制的和金属制的。其中，最受文人雅客和收藏爱好者青睐的是木制笔筒。

Pen Container Bonsai

Pen container is a study appliance. In ancient times, pen container was exclusively used to keep writing brushes. Generally the material to make pen container includes wood, jade, stone, porcelain and metal. Among all types of pen containers, wooden pen containers are most favored by scholars and collectors.

可以选用的植物材料有：五针松、黑松、小叶罗汉松、真柏、榔榆、雀梅、常春藤、迎春、金雀、黄杨、扶芳藤、爬山虎等。选用的器皿可以是：花瓶、茶壶、壶盖、笔筒等实用器皿。

　　异型盆景的造型讲究顺其自然，只要根据植物材料和器皿的特点因势利导就可以了，不需要遵循任何的程式。制作时，植物与器皿要做好大小匹配，比例要相称。如果植物过大，而器皿过小，会让人产生头重脚轻的感觉；如果植物太小，器皿过大，则显得主次不分、喧宾夺主。

is imbued with a unique charm.

　　Generally the creator keeps one's fancy bonsai in small size and commonly chosen plants include *Pinus sibirica*, black pine, shortleaf podocarp, *Juniperus chinensis*, Chinese elm, *Sageratia theezans*, common ivy, winter jasmine, cytisus, Chinese boxwood, wintercreeper, creeper, etc. The containers for fancy bonsai can be household ware such as vase, teapot, teapot top, pen container, etc.

　　Fancy bonsai is styled in a natural and free way. Simply making use of the characteristics of the selected plants and containers is enough and there is no need to follow a certain pattern. When creating fancy bonsai, the grower needs to proportionally select the plants and containers. If the plant is too big for the container, the bonsai will be heavy headed; if the plant is too small for the container, viewers' attention will be shifted from the plant to the container that should not be the focus.

多彩的盆景流派
Colourful Bonsai Genres

在漫长的发展历程中，由于各地的自然条件、文化传统、风土人情和欣赏习惯各有不同，在这些因素的影响下，盆景的造型形式异彩纷呈，加工技艺也更加富有特色。最著名的有八大盆景流派——岭南派、川派、扬派、苏派、通派、海派、浙派和徽派。

In the development of a long history, under the influence of different factors, including various natural conditions, cultural traditions, local customs and practices and habits of appreciation in different areas of China, the style of bonsai also varies and the processing techniques are full of original characteristics. Thus, different genres are brought forth, among which the most famous should be the eight major genres of Chinese bonsai: Lingnan style bonsai, Sichuan style bonsai, Yangzhou style bonsai, Suzhou style bonsai, Nantong style bonsai, Shanghai style bonsai, Zhejiang style bonsai and Huizhou style bonsai.

> 岭南盆景

岭南盆景，又称"广州盆景"，因广州地处五岭之南，故称为岭南。这里气候温暖，日照充足，雨水充沛，草木滋润，得天独厚的自然环境为盆景艺术的繁荣提供了极为有利的条件。盆景制作往往就地取材，选用亚热带和热带常绿细叶树种，如九里香、榕树、福建茶、水松、龙柏、榆树、满天星、黄杨、罗汉松、雀梅、山橘等。盆景的造型布局既来源于自然，又高于自然，力求自然美与人工美的有机结合，被誉为"活的中国画"。

在制作技艺上，"蓄枝截干"的剪裁技法是岭南盆景著名的技法。所谓"蓄枝截干"，就是等盆景树木的枝干长到一定的粗度后进

> Lingnan Style Bonsai

Lingnan style bonsai is also known as Guangzhou style bonsai. Since Guangzhou City is located in the south of the Five Ridges, it is also called Lingnan (south of the ridges). The favourable and special natural conditions in Guangzhou City, including its warm climate, adequate sunshine and rainfall as well as well-grown vegetation, are extremely beneficial for the prosperity of bonsai. To create a bonsai needs to make use of local materials like subtropical or tropical fine-leaved evergreen tree species, including *Murraya paniculata*, banyan, Fujian tea tree, Chinese cypress, juniper, elm, babysbreath, boxwood, podocarp, *Sageretia theezans*, *Citrus japonica*, etc. The inspiration of bonsai's overall arrangement is from and also beyond nature. Lingnan style bonsai tries to integrate the natural and artificial beauty,

• 岭南盆景

当观赏者在一盆岭南盆景前流连时，首先会感觉到它是一棵生长在自然界的古树，是一双神奇的手把它缩小在盆钵之中。然而，经过仔细观赏又会发现，它的一枝一托都分布得巧妙自然，均匀合理，毫无矫揉造作之感。透过这完美的整体造型，观赏者会陶醉在这种天然美之中。

Lingnan Style Bonsai

When admiring a Lingnan style bonsai, firstly people will feel that it is just like an old tree growing in wild and natural conditions. It is the magic craftsmanship that turns it into mini in the container. However, after careful observation and appreciation, it is not hard to find out that every branch is arranged in an ingenious and natural way. The entire work and shape is well balanced. It is such a natural product which is hard to be imaged to be a manmade work. People will definitely be fascinated by Lingnan style bonsai's natural beauty through appreciating the perfect profile of a bonsai.

"蓄枝截干"形成的树冠 （图片提供：FOTOE）

"蓄枝截干"形成的树冠，形如鸡爪、鹿角，佶曲嶙峋。

Crown Based on the Technique "Cutting the Trunk and Growing Branches"

Through using the technique "cutting the trunk and growing branches", the crown will form, whose curved profile is like the chicken feet or deer horn.

行强行剪截，在枝干上选留合适的位置蓄养横枝；待这些新枝干蓄养到适当的粗度后再进行剪裁，使横枝再长出新枝。这样年复一年地再蓄枝、再截枝，逐渐形成树冠。采用此法创作出来的作品，从树干到枝条都能一节一节地按比例缩小，每一节的弯曲角度随由人意，自然流畅，抑扬顿挫，构成苍劲老辣或飘逸潇洒的各种形态，达到缩龙成寸的效果。

"脱衣换锦"也是岭南盆景制作技法上的一大特色。所谓的"脱衣换锦"，又称"脱衣换骨"，是

enjoying the reputation of living Chinese paintings.

With regard to bonsai's manufacturing techniques, the most famous in Lingnan style bonsai should be the pruning technique, known as the "cutting the trunk and growing branches": Cut off the trunk when it grows to certain roughness, and choose an appropriate place on the trunk to cultivate new branches; after those new branches grow to proper roughness, cut them off again and grow new branches. After several years' cultivation and pruning, the crown of the bonsai will form. Through using this method, from the trunk to branches, the whole bonsai can be minimised proportionally and each curve can be controlled. Its sophisticated style is old and vigorous, nature and smooth, just like a loong coiling inside container.

Another special technique used in Lingnan style bonsai is called "changing

指在生长季节对落叶树和常绿树实施人工剪叶，把整株树桩的叶片摘去之后，仍保持树形的优美和自然风貌。岭南盆景讲求古朴自然，刚劲有力，在脱衣换锦之后，毫无矫揉造作之态，更能显示出盆树的骨干苍劲和枝托间的流畅自然。

clothes to gold" or "changing clothes to bone", referring to that bonsai artists defoliate all the leaves of the deciduous or evergreen trees during the growing season, in order to maintain the elegant and natural style of the trees. Lingnan style bonsai pursues the antique,

- 利用"脱衣换锦"技艺制作的岭南盆景
  Lingnan Style Bonsai Manufactured by the Technique "Changing Clothes to Gold"

佛山盆景

在中国盆景中，佛山盆景独具特色，自成一体，是岭南盆景的代表流派。佛山也是岭南盆景的发源地。佛山盆景兴盛于清朝中叶，当时，佛山书画家梁蔼如、梁九图等在自己的私家园林中创建了"十二石斋"，精心制作山石盆景，被后世称为佛山梁园的最大特色。

现在，佛山一共有各类盆景协会十二个，会员多达上千人，佛山盆景已经成为一个新兴的文化产业。有些村镇甚至还将盆景作为一个新型的产业来发展，创出了名气，如顺德陈村和南海平洲。

Foshan Bonsai

Foshan bonsai, a quite unique bonsai school in China, is the representative genre of Lingnan style bonsai while Foshan is the source of Lingnan style bonsai. Foshan bonsai had been prosperous during the middle period of the Qing Dynasty. At that time, Foshan calligraphers and traditional Chinese painters Liang Airu and Liang Jiutu established the Twelve Stones House in the private garden, in order to concentrate on creating rock bonsai, known as the most famous attraction of Liang's Garden in Foshan.

At present, there are totally 12 bonsai associations in Foshan, containing thousands of members. Foshan bonsai has already become an emerging cultural industry. Some towns and villages in this region have developed bonsai as an industry and received reputation and fame, for instance, Chencun in Shunde District and Pingzhou in Nanhai District.

● 树桩盆景
Tree Stump Bonsai

• 树桩盆景
Tree Stump Bonsai

岭南盆景艺术人才辈出，孔泰初、素仁、莫珉府制作的盆景分别代表了岭南盆景的不同风格。

孔泰初（1903—1985）是著名盆景大师，从事盆景创作六十余年，创作出了雄伟苍劲的"大树型"盆景，并首创"蓄枝截干"的

unsophisticated and vigorous profile. Thanks to this method, Lingnan style bonsai is far from artificial, which can better present the balanced and natural relation between the powerful trunk and branches.

There are many bonsai artists in Lingnan, including Kong Taichu, Suren and Mo Minfu, who represent different styles of Lingnan style bonsai respectively.

Kong Taichu (1903-1985), famous bonsai artist, devoted himself to bonsai creation for more than 60 years and created the old and vigorous "large tree" type bonsai. He is also the inventor of the technique "cutting the trunk and growing branches": The trunk is tortuous and old with a prosperous crown. The branches of the bonsai are performed in appropriate thickness.

The artistic style of Suren's (1894-1962) creations shares more characteristics with the freehand brushwork in traditional Chinese painting. With only simple strokes of

▸ 水蜡盆景
Ligustrum obtusifolium Bonsai

技法：树干嶙峋苍劲，树冠丰满，枝条疏密有致。

素仁（1894—1962）所创作的盆景艺术风格更像中国画中的写意派，只要用淡墨描出几笔，便创造出无穷意境。他会根据树木的自然形态来决定最终的姿态，不刻意追求枝繁叶茂，仅仅依靠几枝树丫就能够形成直立挺拔的"高耸型"盆景，效果清高脱俗、如诗如画。

莫珉府（1903—1985），是"自然型"盆景的创建者。他善于借鉴国画构图的思路来创作盆景，构图活泼多样，对比鲜明，别具一格。

light ink, the infinite artistic conception can be created. He would train the bonsai based on the tree's original style, without pursuing luxuriant foliage deliberately. Only depending on several branches, the tall and straight towering tree style bonsai can be created. The effect of this kind of bonsai is free from vulgarity, performing as the scenery in poems and paintings.

Mo Minfu (1903-1985) is the founder of natural style bonsai. He was good at creating the bonsai with the composition conception of traditional Chinese paintings, pursuing the vivid and various artistic effects. This kind of bonsai is quite interesting and distinct, owning a style of its own.

> 川派盆景

　　川派盆景，是以四川命名的盆景艺术流派。根据地域的不同，川派盆景又可以分为川西与川东两种地方风格。川西以成都为中心，包

> Sichuan Style Bonsai

Sichuan style bonsai is a bonsai school from Sichuan Province. In terms of different regions in Sichuan Province, Sichuan style bonsai can be divided into western and eastern Sichuan styles. Chengdu is the centre of western Sichuan, including Wenjiang, Pidu, Xindu districts of Chengdu City, Dujiangyan City, Chongzhou City and Shifang City. Western Sichuan style bonsai pursues the spirit of freehand drawing employed in traditional Chinese paintings, which is relatively delicate and pretty; the central part of eastern Sichuan is Chongqing City, including Chongqing's surrounding

• 川派盆景
Sichuan Style Bonsai

括成都市温江区、成都市郫都区、都江堰市、崇州市、成都市新都区、什邡市等地，在风格上讲究写意，比较清秀；川东则以重庆为中心，包括重庆周围各县，在风格上注重写实，显得浑厚。

川派盆景以桩头盆景而著称。所谓桩头盆景，就是以孤树为主体、没有配景的一种盆景。制作

counties, whose style is realistic, profound and sincere.

Sichuan style bonsai is famous for the solo stump bonsai. The so-called solo stump bonsai refers to the bonsai whose main part is simply a tree stump without any objective view. To create this bonsai, people need to use palm fibre strings to wire the trunk, focusing on the styling method (*Shenfa*). Through the

• **川派桩头盆景**

川派盆景悬根露爪、主干弯曲，树干、树叶、花果古朴而秀雅，风韵苍劲而健茂，给人以自然美和艺术美的享受。

Sichuan Solo Stump Bonsai

The exposed roots of the tree in Sichuan style bonsai resemble loong claws, in which the curved main trunk, together with the branches, flowers and fruits are natural, unsophisticated and delicate. This type of bonsai is famous for its gorgeous type and image. It is old and luxuriant, bringing people the enjoyment of both natural and artistic beauty.

时，棕丝蟠扎，讲究身法，要求古朴严谨，形态弯曲，体态多姿。经常选用的树种有金弹子、六月雪、罗汉松、银杏、紫薇、海棠、梅花、茶花、杜鹃等。

讲究"身法"，是川派盆景在造型上的一大特色。所谓"身法"，是指蟠扎主干的造型方法，在川派盆景中，不同的造型有不同的"身法"。著名的有三弯九倒拐

curved and gorgeous profile, the bonsai is shown in an unsophisticated and antique way. Normally, *Diospyros armata*, *Serissa japonica*, podocarp, ginkgo, *Lagerstroemia indica*, crabapple, plum tree, camellia and azalea are chosen for this kind of bonsai.

Concentrating on style (*Shenfa*) should be a major characteristic of Sichuan style bonsai. The so-called *Shenfa* refers to a method used to wire and train the main trunk. In Sichuan style bonsai, different styles have different *Shenfa*, among which the most famous style is "three curves with paired branches" which means that from the base to the top, nine branches are trained from the main trunk. And for the main trunk, three curves are trained. When admiring the bonsai, people will find nine small curves from the side and three large curves from the front, known as the

• 榔榆盆景
Chinese Elm (*Ulmus parvifolia*) Bonsai

法，即在蟠扎时，从茎基部起向上，蟠扎九个弯子至顶端。然后，在与九拐垂直的立面上蟠曲三个大弯子，三个大弯子形成正面，九个小弯子是侧面，三弯九倒拐由此而得名。

还有方拐法，方拐法即形成方格状的弯子。蟠缚时，先在主干两侧各立一根小竹竿，再扎上一根横的小竹竿，扎成方格形。等到嫩梢长到适当的时候，将嫩梢捆缚到方格上，并使嫩梢的转角成为直角，形成方格状的弯子。这种盆景的成型需要10—20年时间。形态别具一格，给人以新的艺术感受。

川派山水盆景多以崇山峻岭、幽峡奇峰、层岩叠峰、激流险滩等景观为主要题材。由于地方山水环境的熏陶和地方盆石种类资源的影响，川派山水盆景则会使用一些沙片石、钟乳石、云母石、砂积石、龟纹石等为制作石材，在造型上形成了高、悬、陡、深、奇的艺术特征。

reason behind this name.

There is another technique named square turns, referring to that the plants are twisted into square curves. When wiring the tree, firstly two small bamboo poles are placed by both sides of the main trunk, added with a pole above the two poles, in order to form a square framework. When the branches grow after a period of time, they will be bound to the framework and trained into square shape as the corner bends. It takes approximately 10 to 20 years to form this shape, which conduct a quite unique and new artistic effect.

The main themes of Sichuan style bonsai are high mountains and lofty hills, gorgeous valley and different rocks as well as turbulent rivers and treacherous shoals. Influenced by the local landscapes and rock resources, it is quite common for Sichuan style bonsai to employ rocks like seedstone, stalactite, micaalba, sedimentary sandstone and moiré stone and its style is tall, suspended, steep, deep and marvellous.

盆景之乡

郫都区是四川省成都市下辖的一个区，该地区的盆景有着悠久的历史。唐代时该地区的盆景便已进入宫廷，时称"剑南盆景"，后来经过宋、元、明、清历代人士的发扬，该地区的盆景得到了极大的发展。

该地区的盆景造形美观，内涵丰富，制作考究，富含诗情画意，深受人们的喜欢。现在，郫都区的花卉业充分发挥其川派盆景发源地的优势，生产的川派盆景在国内外享有盛誉。

The Home of Bonsai

Pidu is one of the administrative districts of Chengdu city in Sichuan Province. Pidu owns a long history of bonsai creation. In the Tang Dynasty (618-907), bonsai in that area has already been sent to the royal court, known as the Jiannan Bonsai at that time. Under the promotion of professional personnel of the Song (960-1279), Yuan (1206-1368), Ming (1368-1644) and Qing (1616-1911) dynasties, bonsai of Pidu has been improved significantly.

Bonsai of Pidu has gorgeous shapes and profound connotation, which takes exquisite work to create. It shares the characteristics with Chinese classical poetry, enjoying a high popularity. Nowadays, the prosperity of Pidu floriculture industry heavily depends on the advantages of being the home of Sichuan style bonsai, the bonsai produced in Sichuan have been enjoying good reputation at home and abroad.

• 五针松盆景
Pinus sibirica Bonsai

> 扬派盆景

扬州盆景，以江苏扬州为中心，主要包括泰州、兴化、高邮等地。扬州是一座具有两千多年历史的文化名城，处于长江和京杭大运河的交汇处，交通十分发达，加

- 扬派盆景
Yangzhou Style Bonsai

> Yangzhou Style Bonsai

Yangzhou style bonsai is from Yangzhou City in Jiangsu Province, mainly including areas like Taizhou, Xinghua and Gaoyou. Yangzhou is a cultural city, owing more than 2000 years' history and located on the junction of the Yangtze River and the Beijing-Hangzhou Grand Canal. The traffic condition of Yangzhou is wonderful and at the same time, Yangzhou is a fertile place, with a pleasant weather. Early in the Tang Dynasty (618-907), Yangzhou has already been one of the most prosperous commercial cities. Yangzhou has beautiful mountains and clear waters, with exquisite scenery, which have long been the gathering place

上气候宜人、物产丰富，早在唐朝就已经成了全国最繁华的商业城市之一。这里山明水秀，风光旖旎，是文人、商贾的聚集地，在这种环境中孕育出来的盆景艺术，既有北方雄健的特点，又有南方秀美的特征——严整而富有变化，清秀而不失壮观。扬派盆景常用的树种有：松、柏、榆、黄杨、五针松、罗汉松等。制作的时候，棕丝蟠扎，力求精扎精剪。传统造型有游龙弯、云片式和疙瘩式三种。代表人物是万觐棠和王寿山。

of the intellectual as well as businessmen. Yangzhou style bonsai was born in such kind of environment. Therefore, Yangzhou style bonsai combines the vigorous characteristic of Northern China with the delicate feature of Southern China. Generally, Yangzhou style bonsai is neat and changeful, delicate and grand. The generally used tree species in Yangzhou style bonsai are pine, cypress, elm, boxwood, *Pinus sibirica*, and *Podocarpus macrophyllus*. To create Yangzhou style bonsai, people need to use the metal wire to shape the trunk and prune and train the tree in an elaborate way. The traditional styles of Yangzhou style bonsai are roaming loong curves, cloud style and knotty trunk style. The representatives of Yangzhou style bonsai are Wan Jintang and Wang Shoushan.

万觐棠

　　扬派盆景大师万觐棠（1904—1986），江苏泰州人，出身于盆景世家。其先祖万陆奎在清代乾隆年间就曾在泰州城内购地建园，莳花养草，剪扎盆景。之后其子孙承其家业，盆景制作技艺不断完善。清代光绪年间，万氏第四代传人万阳春被召进宫专司盆景，后因其技艺高超，作品精湛，被慈禧太后赐以官职。此后，万家制作的盆景闻名天下。

　　万觐棠少年时即随其家人学习盆景剪扎技艺，成年后不断与同行交流技艺，又融入祖传剪扎技艺之中，形成其独特的艺术风格。

Wan Jintang

Wan Jintang (1904-1986), the representative artist of Yangzhou style bonsai, was from Taizhou City in Jiangsu Province. His family was old and well-known for bonsai creation. His ancestor Wan Lukui had bought land to establish his private garden in Taizhou during the Qianlong Period (1736-1795) in the Qing Dynasty, in which Wan Lukui cultivated flowers and plants as well as created and trained bonsai. His descendants inherited the family business and the bonsai techniques were further perfected. During the Guangxu Period (1875-1908) in the Qing Dynasty, Wan Yangchun, the fourth generation of the Family Wan, was employed by the royal family to administer and manage bonsai specially. Because of his outstanding bonsai techniques and works, Empress Dowager Cixi honoured him with an official position. From then on, the bonsai made by the Family Wan has long been famous in China.

Wan Jintang learned the pruning and wiring techniques of bonsai from his family members when he was a young boy. When he grew up, he exchanged bonsai techniques with other craftsmen. At the same time, he integrated ancestral pruning techniques with the existing bonsai techniques and formed a unique artistic style.

- 万觐棠创作的盆景作品（图片提供：FOTOE）
Bonsai Made by Wan Jintang

扬州盆景以"精扎细剪刀"的剪扎艺术而闻名。"片薄如云"的云片造型和"一寸三弯"的棕扎技艺形成了极富地方特色的盆景风格。

云片是采用棕丝扎法，将各级枝条蟠扎在一个水平面上，形成多种形状的薄片。薄片平而仰卧，似飘浮在树干上的片片云彩，因此被

- 云片

  云片吸取了中国山水画中松、柏远景绘法的经验，不求细节描绘，而是注重画面总体的构图形象，给人一种清秀、宁静、新奇的感觉。

  Cloud Layers

  The creation of cloud layers absorbs the experience used in perspective drawing of pines and cypresses in traditional Chinese painting. It does not pursue the details, but focuses on the general composition of a picture, presenting in a delicate, serene and fresh way.

Yangzhou style bonsai is famous for its pruning and wiring techniques, named delicate and fine pruning. "Delicate cloud style" and "three curves per *Cun* (3.3 cm)" are two representative styles of Yangzhou style bonsai, obtaining the local artistic characteristics.

The cloud layer is manufactured through the wiring technique with palm fibre strings. Branches of each level are trained to the same horizontal plane, in order to form thin layers of different shapes, which are placed smoothly and horizontally, just like the cloud layers floating around the trunk and known as cloud layers. Generally, the foliage on the top is in round shape and foliage in the middle and lower level is generally in palm or fan shape. The size and amount of cloud layers on the main trunk are decided by the size and style of the tree. The most obvious characteristic of the

"一寸三弯"造型的盆景
Three Curves Per *Cun* Style Bonsai

称为"云片"。一般顶部云片形状为圆形，中下部多为掌形或扇形，主干上所分布的云片大小及多少，可以根据植株的大小、树形决定。云片最显著的特点是水平，据说如将盛满水的碟子置于云片之上，也能做到滴水不漏。

"一寸三弯"是根据中国古代画理"枝无寸直"的原则，用棕榈树皮纤维加工成各种粗细棕丝，使寸长之枝也能有三弯。这种融合书法、力学、美学等于一身的蟠扎方法，极大提高了扬派盆景的审美价值。

cloud layer is that they are all horizontally placed. It is said that, a small dish filled with water can be put on the cloud layer without one drop of water spilling.

"Three curves per *Cun* (3.3 cm)" is in accordance with the painting theory that branches should not be straight within one *Cun* (traditional Chinese length unit, one *Cun* equals to 3.3 cm). Through using the palm fibre strings of different degrees of thickness, there will be three curves within one *Cun* of branch. This kind of wiring technique contains many cultural factors of calligraphy, mechanics and aesthetics, which raises the aesthetic value of Yangzhou style bonsai significantly.

扬州盆景园

扬州盆景园位于江苏省扬州市维扬区大虹桥路，属于扬州瘦西湖旅游景点之一，它由"卷石洞天""西园曲水"两部分组成。园内不仅有现代盆景，还现存古盆景四十多盆，其中黄杨盆景"腾云"，曾经在国际大赛中获得金奖。园内除树桩盆景外，还陈列大量水石盆景，在全国盆景界享有极高声誉。

Yangzhou Bonsai Garden

Yangzhou Bonsai Garden, as one of the tourist attractions in Slender West Lake scenic area, is located at Dahongqiao Road, Weiyang District in Yangzhou City. It consists of two parts:

Volume Caves (*Juanshi Dongtian*) and Winding Creek of West Garden (*Xiyuan Qushui*). The garden contains modern bonsai works as well as more than 40 pieces of ancient bonsai. The most famous one among all is the boxwood bonsai Cloud Rider, the winner of an international bonsai competition. Not limited to stump bonsai, there are also many landscape bonsai in this garden, enjoying a high reputation in Chinese bonsai community.

● 扬州盆景园内的盆景
Bonsai in Yangzhou Bonsai Garden

> 苏派盆景

苏派盆景是以江苏苏州为中心的盆景艺术。苏州地处温带，气候温暖湿润，可以供盆景制作的植物品种极其丰富。同时，苏州还盛产太湖石、昆山石、钟乳石、沙积石

> Suzhou Style Bonsai

Suzhou style bonsai was born in Suzhou City, Jiangsu Province. Suzhou is located in the temperate zone and its warm and humid climate allows various species of plants cultivated here for the bonsai creation. At the same time, Suzhou abounds in Taihu stones, Kunshan stones, stalactites and sedimentary sandstones, providing abundant materials for bonsai creation. What is more, since ancient times, Suzhou has attracted many scholars, enjoying strong cultural deposits. Famous writers and poets in different dynasties have left large amount of literatures and

- 苏派盆景
  Suzhou Style Bonsai

等，这些都为盆景制作提供了充分的物质条件。而且自古以来，苏州一直是文人聚集的地方，有着深厚的文化底蕴。历代文人留下了大量的诗文作品，成为盆景艺术家立意的参照。历代画家们笔下的画作也成为苏州盆景中模仿的主题。正是在这多重因素的影响下，苏派盆景显示出独树一帜的风格。

poems, from which many bonsai artists have found themes to create bonsai. Many traditional Chinese painters' works have long been the inspiration for Suzhou style bonsai creation. Just based on these factors, Suzhou style bonsai has shown its own style.

苏州慕园

慕园位于苏州古城人民路和富仁坊的交界处。据说该园原为太平天国时慕王谭绍光的王府，故称为"慕园"。后来，慕园经过修整后改建为专类盆景园。该园占地面积约一千平方米，是中国最早的专类盆景园。园内培植和陈列各类苏派盆景精品，数量达千余盆。圆柏"秦汉遗韵"为该园的镇园之宝，是苏派盆景中的老寿星。树桩使用的是有五百年树龄的圆柏，莲花古盆产自明代，几架则是元代制作的青石古墩，图案由九只狮子组成，人们称之为"九狮墩"。这件珍品将古盆、古树、狮墩巧妙地融为一体，相互辉映，巧妙绝伦。盆中古柏苍劲健茂，树干虽然不高，却大有顶天立地的磅礴气势；树干虽然枯老，枝叶依然青翠，生机勃勃。此盆景构图简洁，虽疏不散，似虚而实，意韵生动。

Suzhou Muyuan Garden

Muyuan garden is located on the juncture of Furenfang Alley and Renmin Road in the old city of Suzhou. It is said that it was the palace of Tan Shaoguang, who was Lord Mu of the Taiping Heavenly Kingdom (1851-1864). It was called Muyuan Garden. Later, Muyuan Garden was restored and rebuilt as a garden especially for exhibiting different kinds of bonsai. Muyuan Garden occupies approximately 1000 m^2, known as the earliest specialised bonsai garden. This garden cultivates and exhibits thousands of various selected works of Suzhou style bonsai. The

"Artistic Spirit of the Qin and Han Dynasties", a Chinese Juniper (*Sabina chinensis*) bonsai, is the treasure of Muyuan Garden and also enjoyed a longevity in Suzhou style bonsai. The trunk used in this bonsai is Chinese Juniper of 500 years' old. The ancient lotus basin was produced in the Ming Dynasty and the ancient stand is made out of bluestone, produced in the Yuan Dynasty. The ancient stand is decorated with the figure of nine lion heads, also known as the Nine-lion Stand. This precious bonsai integrates factors including the ancient basin, old tree and the delicate stand with lion figures, ensuring the balance of these factors and enhancing each other's beauty, which is ingeniously designed. The old Chinese Juniper in the basin is robust and luxuriant. Although the trunk is not tall enough, the whole tree is with great momentum as if it can reach the sky; the trunk is old, but the branches and leaves are verdant, with strong vitality. The composition of this bonsai is simple and clean, sparse but not in random. It is such a perfect piece of art that it is difficult for people to tell whether this is a mirage or not.

● 树桩盆景
Tree Stump Bonsai

"粗扎细剪"是苏派盆景独特的剪裁技法。在蟠扎过程中,苏派盆景力求顺乎自然。在保持其自然形态的前提下,蟠扎其部分枝条,使其枝叶分布均匀、高低有致。其修剪也以保持形态美观、自然为原则,只剪除或摘除部分"冒尖"的嫩梢。

"六台三托一顶"的造型是苏派盆景的特点之一。"六台三托一顶",即将树干蟠成六个弯,在每

"Coarse wiring and fine pruning" is a unique pruning technique in Suzhou style bonsai. For the wiring process, Suzhou style bonsai pursues a way to present the bonsai in a natural way. Under the precondition of maintaining the naturalness, a part of branches are wired in order to make the foliage distribution balanced and well-proportioned. The pruning process is mainly to maintain the beautiful style and the original style of the tree, with only pruning or removing a part of the "cropping up" tender tips.

"Six platforms, three bases and one top" is a unique style in Suzhou style bonsai. "Six platforms, three bases and one top" refers to that the trunk is twisted into six curves, and for each curve one branch is cultivated on one direction. Since one curve has three directions (left, right and back), there are totally three branches for one curve. Namely, totally nine branches are trained for the three curves, among which the six branches in bilateral symmetry are called "six platforms" and the rest three on the back called "three bases". At last, the foliage cluster is wired on the top, known as "one top". This style is interestingly-proportioned and well-arranged. The commonly used tree species are pine,

• 六台三托一顶式盆景
"Six Platforms, Three Bases and One Top" Style Bonsai

个弯的部位留一侧枝,左、右、背三个方向各三枝,扎成九个圆形枝片,左右对称的六片即"六台",背面的三片即"三托",然后在树顶扎成一个大枝片,即"一顶",参差有趣,层次分明。常用的树种有:松、柏、雀梅、榔榆、黄杨等,需要花费十年以上的时间才能完成。

受苏州园林中造园技法的影响,苏派盆景追求情景交融的艺

cypress, *Sageretia theezans*, Chinese elm and boxwood. It takes more than a decade to finish one such kind of bonsai.

Influenced by the techniques used in Suzhou gardens, Suzhou style bonsai pursue the artistic realm of integrating the natural scenery and human emotion. Through using the artistic skill "a miniature of a world", those old and coiled trunks and branches, aged from several decades to hundreds of years, can be trained and grow in a limited

● 室外陈列的盆景
Bonsai Exhibited Outdoors

术境界，运用"小中见大"的艺术手法，将有着几十年乃至上百年的虬干老枝培植于长不盈尺的钵盆之中，再配上古朴的几架，一盆古韵古香的苏派盆景便展现在了世人面前。

container whose length is less than one *Chi* (traditional Chinese length unit, one *Chi* equals to 0.33 m). Equipped with an antique and simple stand, a majestic and delicate Suzhou style bonsai can be presented to people.

苏派盆景大师周瘦鹃、朱子安

周瘦鹃（1895—1968），江苏苏州人，著名作家，苏派盆景大师。早年曾经在上海报馆担任编辑，写作之余喜爱玩弄盆景。晚年致力于盆景的加工、制作。1989年9月，建设部授予其"中国盆景艺术大师"的荣誉称号。他在吸取明清苏州盆景艺术精华的基础上，对苏派的传统技法进行了改革和创新，采用以剪为主、以扎为辅，以及"粗扎细剪"的方法制作树桩盆景，使苏派盆景的面目焕然一新。他制作的树桩盆景清秀古雅，独树一帜。此外，他还撰写了介绍盆景历史和制作方法的专著《盆栽趣味》。

朱子安（1902—1996），江苏常熟人，苏派盆景代表人物。他首创"粗扎细剪"的技法，能使盆景快速定型。他制作的盆景结顶自然，减少了人工痕迹，千姿百态，具有苏州盆景古朴典雅、苍劲自然的风格。

● 树桩盆景
Tree Stump Bonsai

Suzhou Style Bonsai Masters Zhou Shoujuan and Zhu Zi'an

Zhou Shoujuan (1895-1968), from Suzhou City, Jiangsu Province, was a famous author as well as Suzhou style bonsai master. He used to be an editor of a newspaper in Shanghai during his early years. In the spare time, he was fond of researching and training bonsai. When he became old, he was committed to manufacturing and creating bonsai. In September 1989, he was awarded the honorary title of "Chinese Bonsai Master" by the Ministry of Construction. He absorbed the artistic essence of Suzhou style bonsai in the Ming (1368-1644) and Qing (1616-1911) dynasties, and reformed and innovated traditional techniques of Suzhou style bonsai, employing the technique "coarse wiring and fine pruning", which is mainly based on pruning and supported by wiring, to create stump bonsai. He totally changed the presence of Suzhou style bonsai. The bonsai made by him are delicate, ancient and special. What is more, he also wrote a book, named *Interests of Bonsai*, to introduce the history of bonsai and its manufacturing techniques.

Zhu Zi'an (1902-1996), from Changshu City, Jiangsu Province, was a representative artist of Suzhou style bonsai. He invented the technique "coarse wiring and fine pruning" in order to ensure the bonsai be shaped in a short period of time. This technique helps cluster on the top naturally and decreases the artificial traces, making sure the bonsai can be illustrated in different shapes with different expressions. It makes the bonsai to be presented in an antique, pure and elegant way, without losing its power and naturalness, as the characteristics of Suzhou style bonsai.

● 树桩盆景
Tree Stump Bonsai

> 通派盆景

　　通派盆景是以江苏南通命名的一种特色盆景。南通位于长江下游北岸，是一座具有一千多年历史的文化名城。这里气候温和，雨量充沛，土壤肥沃，适宜多种植物的生长。通派盆景通常选用尖短叶小的罗汉松（俗称"雀舌罗汉松"）为材料，蟠扎成"两弯半"的格局。所谓的"两弯半"是将主干从基部开始扎成两个弯，在其上再扎半个弯，从而形成一个完美、自然、立体的S曲线。整个造型雄伟、威武，犹如一尊静坐的雄狮，故又称为"狮式盆景"。

　　通派盆景具有体态丰满的特点。一盆完美的传统通派盆景，树冠呈"馒头顶"，枝片呈"鲫鱼背"，背面要比正面更加丰满。姿

> Nantong Style Bonsai

Nantong style bonsai is a special bonsai school from Nantong City in Jiangsu Province. Nantong is a famous historical and cultural city of more than 1000 years' old, located on the north shore of the lower reach of the Yangtze River. It has mild climate, adequate rainfall and rich soil, which all are good conditions for various plants to grow. The short needled and small leaved podocarp is commonly used in Nantong style bonsai and will be wired as the style two semi-bending. Two semi-bending means that the trunk should be trained into two bends from the base. Above the two bends, a semi-bend is trained, in order to create a perfect, natural and vivid S-curve. The whole style is majestic and mighty, just like a sitting lion, which is also known as the "lion type bonsai".

- 通派盆景

在南通当地流传着这样的谚语：左顾右盼两弯半，云头雨脚美人腰。这两句话可以说是对通派盆景形象的概括。

Nantong Style Bonsai

There is a saying prevailing in the local community of Nantong: The two semi-bending is just like a beauty looking around, whose hair is like cloud, feet like rain with a slender waist. This statement can vividly present the profile and characteristics of Nantong style bonsai.

态威武、气势刚劲的通派盆景多作为室外摆设的盆景，放置于门厅左右。盆景数量多为奇数，以三盆、五盆、七盆最为常见。布置时，主干高大粗壮、稍显端直的主树居中，各盆陪衬树按由高到低的顺序，对称陈列于两侧，营造出一种庄重肃穆的气氛。

One characteristic of Nantong style bonsai is its chubby posture. A perfectly-raised traditional Nantong style bonsai should be like this: Its crown is like "steam bun", and the branches and leaves are like the "back of crucian", with a more chubby back compared with its front. Because of the mighty posture as well as the old and strong appearance, Nantong style bonsai are generally exhibited outdoors, at both sides of the hall. Normally, people deal with the bonsai in odd numbers, commonly in three, five or seven. The bonsai with tall, thick and strong trunks as well as a straight posture should be placed in the middle, set off by other bonsai. The bonsai should be arranged on both sides of the main tree in the middle, from tall to short in order to create a solemn atmosphere.

中国花木盆景之都——如皋

如皋位于江苏省南通市,如皋花木盆景的栽培始于北宋,发展于明清,成熟于20世纪末。如皋盆景的特点是左倚右倾,顶部端正,株干古朴,体形挺秀,多姿多彩,气魄雄伟,尤以雀梅、罗汉松最美观。例如北宋教育家胡瑗遗存的珍品——古桧柏盆景"蛟龙穿云",树龄九百余年;"龙腾虎跃"绒针柏盆景,为明代留存珍品,树龄四百余年。水绘园内存有落地盆景——古桧"六朝柏",俗称"六朝松",相传为北宋名臣曾肇亲手所植,树龄八百多年。这些稀世珍品,向世人昭示着如皋盆景的悠久历史和艺术成就。

Hometown of Chinese Flower and Tree Bonsai — Rugao

Rugao is located in Nantong City, Jiangsu Province. The training history of flower bonsai and tree bonsai in Rugao started from the Northern Song Dynasty (960-1127), developed in the Ming (1368-1644) and Qing (1616-1911) dynasties and matured in the end of 20th century. Rugao bonsai are normally shaped in curved line, with the neat top foliage, old and simple trunk,

- 树桩盆景
  Tree Stump Bonsai

● 石榴盆景
Pomegranate Bonsai

delicate posture, which is majestic and mighty as a whole. *Sagaretia theezans* and podocarp are very representative in Rugao bonsai. There is a precious bonsai, used to be kept by Hu Yuan, who was an educator in the Northern Song Dynasty (960-1127). It is an old juniper bonsai and is more than 900 years' old, called as Loong Shuttling the Clouds. Another bonsai is the needle cypress bonsai, known as "Loongs Rising and Tigers Leaping", dated back to the Ming Dynasty (1368-1644), with a history of more than 400 years. In the Shuihui Garden, there is a landing bonsai of an old Chinese tree named as the "Cypress of the Six Dynasties", also known as the "Pine of the Six Dynasties". It is said that it was planted by a famous official Zeng Zhao in the Northern Song Dynasty (960-1127), of more than 800 years' old. All these rare treasures illustrate the long history and artistic achievements of Rugao bonsai.

通派盆景在造型上的禁忌

通派盆景在造型上有一些禁忌需要注意：一忌扁担干，即在主干上，同一点处向两边一字伸展对生枝，犹如挑扁担，这是不可取的。二忌临面干，即在主干的前面保留侧枝会遮蔽主干，影响其整体气势。三忌背尾干，即在主干的背面拖着一根较长的侧枝，就像尾巴一样，往往会破坏整个树木造型。

Taboos in Nantong Style Bonsai

There are some taboos that need to be noticed in training Nantong style bonsai. Firstly, the branches should not be trained as the shoulder pole, referring to training the branch to grow to opposite sides horizontally from the same place of the trunk, like the shoulder pole. This is not

- 圆柏盆景
  Chinese Juniper (*Juniperus chinensis*) Bonsai

acceptable. Secondly, bonsai craftsmen should avoid training the branches on the front side of the trunk. This means the front side of the main trunk should not be cultivated with branches, which will shield the trunk and affect the mightiness of the bonsai. Thirdly, the branches should be avoided growing on the back of the trunk, meaning that it is not acceptable to cultivate a relatively long branch on the back of the bonsai as a tail, which will damage the general styling of the bonsai.

● 树桩盆景
Tree Stump Bonsai

> 海派盆景

海派盆景是一个以上海命名的中国盆景艺术流派，它的分布范围主要是在上海及其周围各县市。上海地处长江下游的三角洲地带，长江由此处入海，气候温和，四季分明，具有海洋性气候的特点。优越的自然条件和经济文化条件都为海派盆景流派的形成提供了便利。由于上海的特殊地理位置和在国内外贸易中的重要地位，逐步形成了一种勇于革新创造、善于吸收新鲜事物的海派文化。在这种文化的熏陶下，上海盆景广泛吸取了国内各主要流派的优点，借鉴了日本及海外盆景的造型技法，创立了海派盆景。

海派盆景不局限于一种风格，不受任何既有程式的限制，

> Shanghai Style Bonsai

Shanghai style bonsai is a bonsai artistic school from Shanghai City and this kind of bonsai mainly originates in Shanghai and the surrounding cities and counties. Shanghai is located in the Yangtze Delta of the lower reaches of Yangtze River. This place is also the estuary of Yangtze River. Shanghai has a mild climate and four distinctive seasons, obtaining the characteristics of marine climate. The outstanding natural conditions as well as its economic and cultural conditions are all beneficial for Shanghai style bonsai. The unique geographic location of Shanghai has laid its significance in both domestic and international trades, and this has helped Shanghai form a creative and innovative culture. Shanghai is good at embracing new ideas and things and under the influence of such cultural environment, Shanghai style bonsai

在布局上非常强调主题性、层次性和多变性,在制作过程中力求体现山林野趣,重视自然界古树的形态和树种的个性。海派盆景的主要造型有微型和自然型,常

has broadly absorbed the advantages of different bonsai schools in China as well as those overseas, using the bonsai styling techniques of Japan and other foreign countries for reference. All those factors have contributed to the emergence of Shanghai style bonsai.

Shanghai style bonsai is never limited to one single style, free from the limitation of any existing model. Its arrangement especially stresses on performing the theme, level and changeability. The bonsai creation process pursues the rustic charm of mountains and forests, and emphasizes the original shape and natural characteristics of an old tree. The main styles of Shanghai style bonsai are miniatures and natural styles. The commonly used trees are pine and cypress, for instance black pine, masson pine, *Pinus sibirica*, juniper, *Juniperus chinensis*, Chinese elm, *Sageretia theezans*, trident maple, etc.

Shanghai style bonsai concentrates on the old trees of various shapes, using the painting techniques in traditional Chinese landscape paintings for reference, training and manufacturing the trees based on the trees' original conditions, which can increase the naturalness

• 海派盆景
Shanghai Style Bonsai

用树种以松柏类为主，如黑松、马尾松、锦松、五针松、桧柏、真柏、榔榆、雀梅、三角枫等。

海派盆景以自然界千姿百态的古木为摹本，参考中国山水画的画树技法，因势利导，进行了创造性的艺术加工，赋予了作品更多的自然之态。海派盆景的造型，形式自由，不拘格律，没有任何固定的程式，追求"自然入画，精巧雄健，明快流畅"的造型风格。造型形式多种多样，主要有直干式、斜干式、曲干式、临水式、悬崖式、枯干式、连根式、附石式等。此外还有一种"点石式"盆景，为了增加山野情趣，会在树木盆景内结合场景的氛围配置一些山石。

of the bonsai. Shanghai style bonsai is free from the limitations of rules, getting rid of any fixed manipulation. It concentrates on performing the bonsai in a "natural, picturesque, delicate and mighty way, whose styling is vivid and smooth". There are different kinds of styling, like straight trunk style, slanting trunk style, curved trunk style, water-leaning style, cliff-hanging style, withered trunk style, connected root style, root-over-rock style. What is more, there is a style called as "spot stone style". In order to increase the rustic charm of mountains and forests, the main tree is decorated with some stones and rocks.

上海植物园盆景园

上海植物园盆景园是国内最大的国家级盆景园，占地五十余亩，园内汇集了数千盆以海派盆景为代表的盆景精品。

该园由树桩展示区、山石盆景展示区、海派盆景博物馆和盆景养护区四个区域组成。树桩盆景区展示了千余盆桩景精品，各式盆景星罗棋布，姿态万千，意境深远。经过几代盆景工作者的创作研究，融汇众家之所长，拜自然为师，形成了自己独特的海派盆景艺术风格，师法自然、苍古入画、意境深远。

Bonsai Garden in Shanghai Botanical Garden

Shanghai Botanical Garden contains the largest national bonsai garden in China. It occupies more than 3 hectares and has thousands of selective Shanghai style bonsai.

The bonsai garden is consisted of four parts: Stump Bonsai Showcase, Rock Bonsai Showcase, Shanghai Bonsai Showcase and Bonsai Maintenance Area. Different kinds and styles of bonsai are illustrated, forming a far-reaching conception. Thanks to the efforts and study of several generations of bonsai creators and craftsmen, this garden has integrated the advantages of different bonsai schools. What is more, through learning from the nature, the unique artistic style of Shanghai bonsai has formed. It presents the nature perfectly with an old, picturesque and far-reaching conception.

上海植物园盆景（图片提供：FOTOE）
Bonsai in Shanghai Botanical Garden

> 浙派盆景

浙派盆景，是以浙江命名的一种盆景流派，主要流传于杭州、温州两地。浙派盆景以松柏为主体，以高干合栽为基调，艺术形象高昂挺拔，遒劲而潇洒，严谨中有舒展，豪放中见优雅。作品讲求造型简洁，注重内涵和气韵。

浙派盆景在造型上以江南古树

> Zhejiang Style Bonsai

Zhejiang style bonsai is from Zhejiang Province, a bonsai school prevailing in Hangzhou City and Wenzhou City. The commonly used tree species in Zhejiang style bonsai are pine and cypress. Zhejiang style bonsai is famous for its tall trunk style, showing its majestic and tall appearance as well as its lean and strong frame. The whole work is vigorous and elegant, without

- 浙派盆景——独钓寒江雪（图片提供：FOTOE）
  Zhejiang Style Bonsai—Fishing Alone in the Cold Snow

为蓝本，主枝粗壮遒劲，线条曲折起伏，伸缩自如，极具变化美。小枝剪成鸡爪鹿角状，枝片也是枝中有枝，片中有片。每枝每片都能独立成景，颇具诗情画意。

在继承浙江地域特色的基础之上，浙江盆景继承了自唐、宋以来在浙江广为流传的"天目石松"那种苍翠挺拔的傲然风骨，以松柏类植物作为浙江盆景制作的主要树

any restrictions. This style pursues a simple profile and concentrates on the connotation and the control of the artistic conception of the bonsai.

Zhejiang style bonsai choose Jiangnan (regions south of the Yangtze River) ancient trees as the styling model, whose main trunk is thick and vigorous, with a curved line, showing its beauty and charms through its changeability in the profile. The branches are pruned into the shapes of chicken feet or deer horn, and are well-trained into distinctly shaped layers. In that way, each branch and piece of leaf can be viewed as poetic scenery.

Zhejiang style bonsai has inherited the regional characteristics and absorbed the majestic and solitary features of the tall staghorn style bonsai, used to be

- **浙派盆景**（图片提供：FOTOE）
浙派盆景严谨、稳健、端庄，枝叶清晰，浑厚华润，具有洒脱奔放的气势。
Zhejiang Style Bonsai
Zhejiang style bonsai are majestic and elegant, with clear layers and distribution of branches and leaves, showing in a magnificent but unrestricted way.

种。在盆景造型上，与传统的人工扭曲的S形相抗衡，崇尚直式、高干的自然姿态，多采用棕丝和金属丝蟠扎与细修精剪相结合的造型技法，逐步形成了有别于江南其他各派的艺术风格。为了表现莽莽丛林的特殊艺术效果，通常都会将直干或三五株栽到一个盆里；为了表现苍古意趣，还会对柏类的主干做适度的扭曲并剥去树皮。

prevailing in the Tang (618-907) and Song dynasties (960-1279). The trees generally used in Zhejiang style bonsai are pine and cypress. Meanwhile, by using the metal wires and palm fibre strings, the wiring technique is generally combined with meticulous pruning, which helps Zhejiang style bonsai form its own special artistic style, differing from other bonsai schools in this area. In order to show a unique artistic effect of a luxuriant forest, normally one straight trunk or even three to five trunks will be cultivated in one container. And in order to better perform the antique trees, artists sometimes will twist the main trunk of cypresses moderately and remove the bark.

盆景的命名

为盆景命名是中国盆景的一大特色。当盆景制作者完成一件作品之后，还要为盆景起一个名称。名称既是盆景的标识，也是盆景灵魂的重要体现。好的盆景名称不但能概括主题，诠释作品内涵，还能引起读者的遐思，给人以丰富的想象。

引用古代诗句是盆景命名中常用的方法之一。如果山水盆景中有山峰耸峙，水中漂浮着一叶扁舟，就可以引用唐代诗人李白《早发白帝城》中的诗句"轻舟已过万重山"命名，形象生动。

以风景名胜命名盆景也很常见。位于广西的桂林山水，山形奇秀，石色苍蓝。如果山水盆景与其有相似之处，就可以此命名。此外，像"泰山神韵""洞庭早春"等等都可以作为盆景的名称。

The Name of Bonsai

The name of bonsai should be another outstanding feature of Chinese bonsai. To finish the creation of bonsai, it is necessary to name the work as well. The name of bonsai should also be a hallmark, an important presentation of the bonsai's soul. A good name should not only refer to the theme of the bonsai, explaining the connotation of this work, but also stimulate people to re-think, leaving people with a space to image.

Quoting the classical Chinese poems should be one of the commonly used methods to name the bonsai. If there are rockeries in the container as the mountains, with a small boat floating on the lake, it is suitable and vivid to use Li Bai's poem "Lightly, our boat has already passed ten thousand green mountains" from *Sailing Early From Baidi Town*.

It is also common to see that the bonsai is named after places of interest. Located in Guangxi, Guilin has delicate and miraculous mountains and blue coloured rocks. If there is any resemblance existing between the bonsai and the scenery, this bonsai can be named after the place. What is more, the Verve of Mount Tai and the Early Spring of Dongting Lake can also be bonsai names.

• 山石盆景
Rock Mountain Bonsai

> 徽派盆景

徽派盆景，是以徽州命名的一种盆景艺术流派。徽派盆景以歙县的卖花渔村为代表，主要包括绩溪、黟县、休宁等地民间制作的盆景。徽州地处新安江上游，气候温和，雨量充沛，适宜花木的生长。兴旺发达的经济和深厚的文化底蕴，为徽派盆景提供了坚实的经济基础和文化环境。

徽派盆景风格独特各异，形式多种多样，造型风格以古傲苍劲、奇峭多姿为主要特色。传统造型主要有游龙式、扭旋式、三台式、屏风式和疙瘩式，技法特点为粗扎粗剪。

徽派盆景植物种类较多，常用树种以梅桩最为著名，称为"徽

> Huizhou Style Bonsai

Huizhou style bonsai is a bonsai school from Huizhou. The representative works of Huizhou style bonsai are from Maihuayu Village in Shexian County, including the bonsai made by ordinary people from counties like Jixi, Yixian and Xiuning. Huizhou is located on the upper reaches of Xin'an River, which has mild climate and abundant rainfall, all presenting as good conditions for the growth of trees and flowers. The prosperous and developed economy and profound culture have provided solid economic and cultural bases for the development of Huizhou style bonsai.

Huizhou bonsai contains a wide range of styles and can be performed in different ways, which are majestic, vigorous and ingenious. The traditional

- **徽派梅花盆景**（图片提供：全景正片）
  古人认为梅花以曲为美，徽州梅花盆景就是以此为依据进行造型的。徽州梅花盆景讲究对称，格调凝重。

  Huizhou Plum Bonsai
  Traditionally, Chinese ancients thought that beautiful plum tree must be in curved shape and this is how Huizhou plum bonsai is trained. Huizhou plum bonsai pursues symmetry, owning a dignified profile.

梅"。其他的如徽柏、黄山松、罗汉松、翠柏、紫薇、南天竹、榔榆等也都比较常见。

徽派盆景在制作上采用"粗扎粗剪"的造型艺术手法。对于幼小的枝条，通常是用棕榈叶片进行定坯造型，每两年重新调整一次；对于较大的枝干改用棕绳

styles include coiled loong style, twisting style, three-platform style, screen style and knotty trunk style. "Coarse wiring and coarse pruning" is a unique pruning technique in Huizhou style bonsai.

Huizhou style bonsai has different categories, among which the most famous is the plum tree bonsai, known as Huizhou Plum Tree. Other commonly

梅花的寓意

在中国，梅花深受人们的欢迎。它清雅俊逸的风度是历代诗人、画家赞美的对象，它迎风冒雪的坚强品格被喻为中华民族的象征。

此外，在中国民间的众多日常生活用品中也可以经常见到梅花的图案。在民间，梅花被认为是传春报喜的吉祥象征。梅花和荷花组合，寓意和和美美。梅花与牡丹组合，寓意富贵眉寿。梅花上落有喜鹊，寓意喜上眉梢。

Implied Meanings of Plum Blossom

Plum blossom is quite popular in China. It is delicate and solitary, enjoying the poets and painters' praise for many dynasties. Plum blossom braves wind and snow, praised as the symbol of Chinese nation's strong character.

Moreover, it is common to see the design of plum blossom on the daily necessities of Chinese people. Plum blossom is regarded as the auspicious symbol, transmitting the messages of spring and good news. The combination of plum blossom and lotus means reunion and happiness. The combination of plum blossom and peony means richness and longevity. The combination of plum blossom and magpie means there will be good news soon.

- 山东潍坊年画《梅花开五福》

Plum Blossom for Five Kinds of Happiness, A New Year Picture of Weifang City, Shandong Province

蟠扎，等到主干大致定形后再对侧枝进行加工，对小枝则只做修剪不做蟠扎。

used species include Huizhou cypress, *Pinus hwangshanensis*, podocarpus macrophyllus, bluish green cypress, *Lagerstroemia indica*, *Nandina domestica* and Chinese elm.

"Coarse wiring and coarse pruning" is commonly used as a styling technique in Huizhou style bonsai. For those shoots, artists usually use palm leaves to train the bonsai. There will be two rearrangements in one year; for those relatively large trunks and branches, the palm fibre strings are employed to wire the bonsai. After the main trunk is generally styled, artists will manufacture the branches. They will only prune the branches without wiring them.

• 梅花盆景
Plum Bonsai

卖花渔村

卖花渔村本名洪岭村，位于安徽歙县县城东南，是中国著名的盆景之乡。这里气候温暖湿润，雨量充足，土壤肥沃，为盆景植物的生长及树桩的培育提供了优越的场所。卖花渔村以产梅为主。在这个村子里，家家户户在山上都辟有花木场、建有盆景园，少则数千株，多则数万株。每年都会将大量的盆景销往上海、北京、武汉、南京、广州等地。

Maihuayu Village

Maihuayu Village also called Hongling Village in the past, located in the southeast of Shexian County in Anhui Province, is a famous source of Chinese bonsai. The climate there is warm and humid, with abundant rainfall and rich soil, providing outstanding conditions for the cultivation and training of bonsai plants and trees. Maihuayu Village is famous for its plum blossom. Approximately every family in this village has its own flower and tree garden as well as bonsai garden, from thousands of to tens of thousands of plum trees. Every year, this village will sell large amounts of bonsai to different places in China including Shanghai, Beijing, Wuhan, Nanjing and Guangzhou.

● 树桩盆景
Tree Stump Bonsai

中国徽派盆景艺术博物馆——歙县鲍家花园

鲍家花园位于安徽省歙县境内，原为清朝著名徽商鲍启运的私家花园，后废于战乱。2000年该园得以复修扩建，收藏了万余盆徽派盆景。其中园内的超大型山水盆景《江山如此多娇》十分引人注目。它长20.0米，宽3.2米，高2.4米，以芦管石为主石料，配以砂积石堆成。整个作品以峰、峦为主调，展现了祖国大好河山的恢弘气势。

Huizhou Style Bonsai Art Museum—Bao's Garden in Shexian County

Bao's Garden is located inside Shexian County in Anhui Province, used to be the private garden of an Anhui businessman whose name was Bao Qiyun in the Qing Dynasty (1616-1911) and was once destroyed in chaos caused by war. In 2000, this garden has been restored and expanded, owning tens of thousands of Huizhou style bonsai. There is a large landscape bonsai named *What A Great China,* which has attracted many people's attention. It is 20.0 m long, 3.2 m wide and 2.4 m tall, using reed tube stones as the main material, decorated with sedimentary sandstone piles. The whole work concentrates on mountains and hills, illustrating the majestic, grand and wonderful rivers and mountains in China.

- 鲍家花园内的《徽州人家》大型山水盆景（图片提供：全景正片）
Large Landscape Bonsai *Huizhou Family* in the Bao's Garden